James Clerk Maxwell's equations for the electromagnetic field provide the basis for a good portion of twentieth-century science and technology. Two of Maxwell's contributions in this area – the displacement current and the electromagnetic theory of light – are among the most spectacular innovations in the history of physics, not only because of their immense impact but also because they constitute paradigmatic examples of theoretically motivated innovation, undertaken on the basis of little or no experimental evidence. Historians and philosophers of science have been drawn to this subject, but the technical complexities and thematic subtleties of Maxwell's work have turned out to be difficult to unravel, and scholarship to date has generated more questions than answers.

The key to an understanding of Maxwell's innovations, according to Daniel Siegel's analysis, has been visible all along, if unappreciated: Interspersed through Maxwell's work in electromagnetic theory are certain mechanical models of the electromagnetic field, which have commonly been regarded as amusing but irrelevant. It becomes clear, however, on the basis of a close analysis of the original texts – with careful attention to the equations as well as the words – that mechanical modeling played a crucial role in Maxwell's initial conceptualizations of the displacement current and the electromagnetic character of light. Dr. Siegel also locates Maxwell's work in the full sweep of nineteenth-century electromagnetic theory – from Oersted, Ampère, and Faraday through Hertz and Lorentz – and in the context of the methodological traditions and perspectives on the nascent physics discipline at the universities of Edinburgh and Cambridge.

Innovation in Maxwell's electromagnetic theory

Innovation in Maxwell's electromagnetic theory

Molecular vortices, displacement current, and light

DANIEL M. SIEGEL
University of Wisconsin

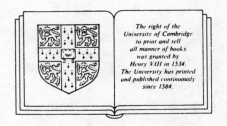

The right of the
University of Cambridge
to print and sell
all manner of books
was granted by
Henry VIII in 1534.
The University has printed
and published continuously
since 1584.

CAMBRIDGE UNIVERSITY PRESS
Cambridge
New York   Port Chester   Melbourne   Sydney

PUBLISHED BY THE PRESS SYNDICATE OF THE UNIVERSITY OF CAMBRIDGE
The Pitt Building, Trumpington Street, Cambridge, United Kingdom

CAMBRIDGE UNIVERSITY PRESS
The Edinburgh Building, Cambridge CB2 2RU, UK
40 West 20th Street, New York NY 10011–4211, USA
477 Williamstown Road, Port Melbourne, VIC 3207, Australia
Ruiz de Alarcón 13, 28014 Madrid, Spain
Dock House, The Waterfront, Cape Town 8001, South Africa

http://www.cambridge.org

First published 1991
First paperback edition 2002

A catalogue record for this book is available from the British Library

Library of Congress Cataloguing-in-Publication Data
Siegel, Daniel M.
Innovation in Maxwell's electromagnetic theory : molecular
vortices, displacement current, and light / Daniel M. Siegel.
p. cm.
ISBN 0 521 35365 3 hardback
1. Electromagnetic theory. 2. Physics – History. I. Title.
QC670.S48 1991
530.1′41–dc20 90-42511
CIP

ISBN 0 521 35365 3 hardback
ISBN 0 521 53329 5 paperback

For the women in my life
Rebecca
Shulamith
Ruth
Rebecca, Deborah, Sarah

CONTENTS

PREFACE

The historian of science cannot be unmindful of the fact that, for better or for worse, science, as practiced now and in the past, furnishes one of our central models of rational thought and judgment. Some would use history to demonstrate the worthiness of science as a model for rationality; others would use history to demonstrate the limitations of science in this respect. Above all, however, awareness of the paradigmatic role of science urges the historian of science to seriousness of purpose in trying to delineate and understand the practice of science in the past. In particular, I here endeavor to delineate and understand, in historical context, James Clerk Maxwell's seminal work in electromagnetic theory. Understanding of matters of any significance, however, seems never to come easily: This book has been long in gestation, and it makes some demands of its reader.

I was introduced to the historical study of Maxwell by Martin Klein, whose work and counsel have been seminal for me.

Jed Buchwald, Francis Everitt, Peter Harman, John Heilbron, Ole Knudsen, and David Wilson have furnished ideas, sources, and encouragement beyond what scholarly citations can acknowledge.

My work has been enriched by the conversation as well as the scholarly publications of Joan Bromberg, Geoffrey Cantor, Alan Chalmers, Michael Crowe, Gregory Good, John Hendry, Jonathan Hodge, Robert Kargon, Donald Moyer, Richard Olson, Paul Theerman, and Norton Wise. Interaction with other participants during the conference "Cambridge Mathematical Physics in the Nineteenth Century," held at Grasmere, England, in March 1984 – which resulted in the volume *Wranglers and Physicists,* edited by P. M. Harman (Manchester University Press, 1985) – opened new vistas.

My colleagues over the years in White Hall at the University of Wisconsin – William Coleman, Victor Hilts, David Lindberg, and

Robert Siegfried – have been supportive intellectually and personally. My students in recent seminars on Maxwell and on scientific method in the nineteenth century, including Michael Boersma and Derrick Mancini, have helped in the clarification of many issues.

Helen Wheeler and Cambridge University Press have given help and encouragement throughout the writing and publishing process.

The Cambridge University Library has been generous with assistance and permissions pertaining to Maxwell materials; I thank also the Edinburgh University Library for permission to quote from Balfour Stewart's notes on James D. Forbes's lectures.

The National Science Foundation and the Graduate School of the University of Wisconsin have been financially supportive.

Much of Chapter 2 appeared in *Wranglers and Physicists* and is used here with the permission of Manchester University Press; much of Chapter 4 appeared in *Historical Studies in the Physical and Biological Sciences, 17* (1986), 99–146, and is used here with the permission of the University of California Press.

INTRODUCTION

James Clerk Maxwell made momentous contributions to the development of electromagnetic theory: In formulating the set of equations that bear his name, he established a systematic and enduring foundation for modern electromagnetic theory; in developing the formalism to embrace optics, he demonstrated the range and power of his mathematized field theory, adumbrating its profound implications for subsequent developments ranging from relativity theory to communications technology. Maxwell's activity in this area spanned a period of twenty-five years – from the mid-1850s until his death in 1879 – and his thinking on the subject was developing and changing throughout that period. It is possible, nevertheless, to identify one crucial period of innovation: a period of about one year, centering on the summer of 1861, during which Maxwell was working on, and publishing in successive installments, a paper entitled "On Physical Lines of Force." It was during that period that Maxwell modified one of the fundamental electromagnetic equations through the introduction of a new term called the displacement current, thereby rendering the set of foundational equations complete and consistent; and it was also during that period, in conjunction with the introduction of the displacement current, that Maxwell took the crucial first steps toward the unification of electromagnetism and optics. These two pivotal innovations – the introduction of the displacement current into the equations and the assimilation of optics into the resulting theory – were undertaken in the context of Maxwell's work on an elaborate mechanical model of the electromagnetic field, involving vortical motions of molecular portions of the "magneto-electric medium," or ether; he called the resulting theory the "theory of molecular vortices." My purpose here is to investigate this nexus of innovation in Maxwell's electromagnetic theory, involving the elaboration of the molecular-vortex model, the introduction of the displacement current, and the incorporation of light into the theory.[1]

Because of the central importance of this episode in the development of modern physics, and also because of its paradigmatic status as an example of theoretically motivated innovation in science, it has been the subject of much attention. Making sense of the history of Maxwell's electromagnetic theory, however, has turned out to be less than straightforward – there are both technical complexities and thematic subtleties that have proved difficult to unravel – and scholarship on the subject to date has generated rather more questions than answers.[2] The perennial issues in Maxwell scholarship to be addressed by this investigation may be phrased in terms of three sets of questions. The first pair of questions relates to Maxwell's extensive use of mechanical models in developing his electromagnetic theory: (a) To what extent did Maxwell believe in these models as representations of reality? (b) To what extent did these mechanical models actually help Maxwell in his work on electromagnetic theory?[3] A second pair of questions relates to the physical conceptions – mechanical or other – that were operative in Maxwell's field theory: (a) How did he conceptualize electric charge, electric current, and the electric and magnetic fields? (b) What picture did he have of the connections and relationships among charges, currents, and fields?[4] Third, there are questions concerning the unity and coherence of Maxwell's mechanical models and mathematical formalisms: (a) Did he make use of a congeries of mutually inconsistent mechanical models, merely for suggestive or illustrative purposes, or did he in fact achieve at some point a consistent and unified mechanical account of electromagnetic phenomena? (b) Are the equations for the electromagnetic field quantities that Maxwell used (which differ in subtle but significant ways from the equations that we now use) consistent and coherent when viewed in proper historical context, or do the discrepancies between the equations that Maxwell wrote and the modern equations (in particular, differences in algebraic signs) point to some fundamental incoherence in Maxwell's original formulations?[5] This study builds on previous discussions of these general questions, but it goes beyond, especially in relating these broader questions to a detailed investigation of the emergence of the displacement current and the electromagnetic theory of light in the context of the molecular-vortex model.

By way of introduction, the broader background for Maxwell's work is sketched in Chapters 1 and 2: Chapter 1 describes the general situation in nineteenth-century electromagnetic theory, as well as the particular scientific training and methodological perspectives that Maxwell brought to the subject. Chapter 2 then surveys Maxwell's changing attitudes toward the use of mechanical models in electromagnetic theory throughout his career. Chapters 3–5 proceed to a detailed account of the emergence of Maxwell's crucial innovations in the context of the molecular-vortex model: Chapter 3 deals with the elaboration of the molecular-vortex model, Chapter 4 with the introduction of the displacement current, and Chapter 5 with the initial formulation of the electromagnetic theory of light. Chapter 6, finally, deals with the further development and subsequent impact of Maxwell's innovations. The sum of these parts does not amount to a complete survey of Maxwell's work in electromagnetic theory; instead, the aim is to concentrate on the innovative nexus and directly relevant contextual material, in order to achieve the depth of analysis required for making progress in addressing the somewhat knotty issues involved.

To anticipate conclusions – in brief and without nuance – with respect to the three major issues delineated: First, as concerns Maxwell's use of mechanical models, (a) he had a stronger ontological commitment to the molecular-vortex model than has generally been appreciated, and (b) that commitment furnished the motivation for Maxwell to dedicate substantial effort to careful articulation of the model, which resulted in a model capable of playing a central and productive role in his thinking about electromagnetic theory.[6] Second, with respect to Maxwell's conceptualization of (a) charges, currents, and fields and (b) the relationship among these, Maxwell was in this matter a devoted follower of Michael Faraday, who did not believe in the existence of primitive charge carriers, and who viewed charges and currents as emergent from fields, rather than vice versa. Maxwell's commitments in this direction were reflected in his initial formulations of the molecular-vortex model, the displacement current, and the electromagnetic theory of light, giving to all of these a conceptual orientation quite different from that characteristic of his own later formulations of electromagnetic theory and their sequelae in twentieth-century classical field theory.[7] Third, taking all of this into account, it becomes possible to make sense of the innovative nexus in Maxwell's electromagnetic theory, and it thereby becomes evident that there is more coherence and consistency in (a) the mechanics and (b) the mathematics of the molecular-vortex model (and hence the whole innovative nexus) than has hitherto been appreciated.

In order to make sense of Maxwell's mathematics, it will be advantageous to develop a uniform mnemonic notation in which to render the various symbolisms that Maxwell used. It is certainly true that greatest fidelity to concrete historical detail would be achieved by using only Maxwell's original notation; there would, however, be two difficulties associated with such a choice – mnemonic and analytical. With respect to mnemonics, the fact that Maxwell used primarily component notation – indeed, a variety of component notations at various times and in various places – would mean that a large number of symbols would have to be kept in mind in order to construct a mental picture of the development of Maxwell's thought; furthermore, confusion would arise from Maxwell's use of the same symbol to represent different things in different places. Clearly, then, there are great mnemonic advantages to be gained from using a uniform vector/tensor notation. In addition, there is an analytical advantage to using uniform notation: Historical analysis often involves comparison and contrast among the various stages of historical development, and comparison and contrast of mathematical forms are much facilitated by reduction to common notation. There are, of course, dangers involved in any translation, including the contemplated translation to unified vector notation. In order to mitigate the dangers of translation, my procedure in critical cases (see especially Chapter 4) will be to begin with the component equations as Maxwell wrote them, thence proceeding to explicit discussion, as appropriate, of the problems and issues involved in their conversion to vector form. Where the conversion to vector form is less problematic (as in Chapter 3), there will be less in the way of explicit discussion of the conversion process. (However, the component symbols that Maxwell used will in all cases be indicated, either in the text or in the notes.) I shall make use of the standard modern scalar, vector, and tensor symbols when the context appears to justify that, or alternative symbols when it appears that Maxwell's quantities cannot be assimilated to the modern ones. The resulting equations will furnish a mnemonic record of the judgments made concerning the similarity or dissimilarity between Maxwell's physical quantities and our modern ones. It is my judgment that the translation of component equations into vector form in this fashion will introduce minimal distortion, while greatly aiding understanding. I shall take pains, however, to avoid any alteration either of algebraic signs or of placement and grouping of terms in the equations, as these will turn out to be of great importance (especially in Chapter 4), and to alter these features of the equations would be to distort their meaning. Decisions concerning the rendering of the equations will thus be made on a heuristic basis – with the aim of facilitating analysis and enhancing understanding – rather than on the basis of any a priori commitment concerning the use or nonuse of modern notation and terminology in historical discourse.[8]

1

The background to Maxwell's electromagnetic theory

The first half of the nineteenth century was marked by the discovery of a variety of new phenomena in electricity and magnetism, and the general task to which Maxwell then addressed himself, beginning in the 1850s, was the development of a unified electromagnetic theory that would incorporate all of these phenomena. Section 1 of this chapter explores the broad background in nineteenth-century electromagnetic experiment and theory for Maxwell's work. Section 2 deals with the particular scientific and mathematical education and training that Maxwell brought to his work in electromagnetic theory, including his educational experiences at the universities of Edinburgh and Cambridge, as well as the epistolary tutelage of William Thomson. Finally, Section 3 deals with the methodological approach – growing out of the combined Scottish and Cambridge backgrounds – that Maxwell brought to his work.

1   *Electricity and magnetism from Oersted to the 1840s*
        The discovery of the magnetic effects of electric currents, announced by Hans Christian Oersted in July 1820, defined the agenda for the study of electricity and magnetism during the following decades. Viewed in broad terms, Oersted's discovery had implications in two directions: First, Oersted's research had been undertaken in the context of widespread preoccupation – following Alessandro Volta's invention of the voltaic pile in 1800 – with the connections between electric currents and other physical phenomena, including chemical effects, heat, and light. Oersted's striking discovery of a relationship between electricity and magnetism furnished strong reinforcement for this concern with the connectedness of physical phenomena in general; the relationship between electricity and magnetism, in particular, soon came to be viewed as paradigmatic. Research in electricity and magnetism after Oersted thus became centrally concerned with exploration of the relationships among

electricity, magnetism, the media sustaining these effects, and a variety of related phenomena. A second implication of Oersted's discovery – as he understood it and as widely perceived – was the existence of a new kind of force: a circular force distributed in space, in contradistinction to the push-pull forces known previously in mechanics, electricity, and magnetism. This aspect of Oersted's discovery fostered continuing concern in the following decades with the directionality and spatial arrangement of electrical and magnetic influences, as well as their mode of transmission through space. The dual agenda for the study of electricity and magnetism resulting from Oersted's work was quite universally accepted. There were, however, differences of opinion as to the way in which that agenda was to be developed, eventuating in the establishment in the 1820s – and persistence through much of the rest of the nineteenth century – of two distinct schools in electromagnetic theory: the Continental school, stemming from the work of André-Marie Ampère, and the British school, developing from the work of Michael Faraday.[1]

Ampère, responding initially to Oersted's discovery in the fall of 1820, set out to mitigate some, but not all, of the revolutionary implications of Oersted's work. Inverse-square central forces were fundamental to the tradition of mathematical physics and astronomy that flourished in France in the late eighteenth and early nineteenth centuries; the application of that tradition to electricity and magnetism was exemplified in the work of Charles Augustin Coulomb and Siméon Denis Poisson. Committed to the central-force foundations of that tradition, and wielding its mathematical methods expertly, Ampère managed to find a way of reducing Oersted's circular forces to the familiar inverse-square central forces. Ampère's theory depended on the assumption of molecular currents – electric currents circulating in the molecular portions of matter. Magnetization was then attributed to preferential orientation of the molecular currents within the magnetic material in a particular direction, namely, with the normals to the planes of the circular currents preferentially oriented along the magnetic axis. The interaction of a magnet with an electric current in a wire could then be described in terms of forces exerted between the molecular currents in the magnet and the macroscopic current in the wire. The form of the current–current interaction was specified, at least in part, on the basis of macroscopic experiments. Ampère's conclusion was that all forces among macroscopic currents and magnetic materials could be accounted for in terms of an interaction between current elements represented by the following equation:

$$F = \frac{ii' \, ds \, ds'}{r^2} \left(\sin \theta \sin \theta' \cos \omega - \frac{1}{2} \cos \theta \cos \theta' \right) \qquad (1.1)$$

where F represents a push-pull force between the current elements (attractive if the numerical value is negative, and repulsive if positive), i and i' represent the intensities of the two currents, ds and ds' represent the lengths of the current elements, r is the distance separating the two current elements, sin and cos represent trigonometric sine and cosine, θ and θ' are the respective angles that the line elements ds and ds' make with the line connecting the current elements, and ω is the angle between the planes defined by the two current elements each taken with the connecting line. Considered in terms of basic essentials, this was an inverse-square push-pull force of familiar type; the angle-dependent factor in the equation, however, constituted a novel complexity.[2]

Maxwell referred to Ampère as the "Newton of electricity," and the parallel between Ampère and Newton extends not only to the mathematical forms of the inverse-square force laws that each proposed (as well as the unifying power of these force laws in celestial mechanics and electromagnetism) but also to their respective opinions concerning the physical basis of such force laws. Thus, whereas both Newton and Ampère were seen by some of their followers as believers in unmediated action at a distance – between mass elements or current elements, respectively – both in fact conceptualized such interactions as mediated by an intervening medium – an "ether." Ampère viewed the etherial substance as arising from the combination of positive and negative electrical fluids, and he considered the ether to be involved in the propagation of optical and thermal effects as well as electrical and magnetic effects. Ampère thus took seriously the imperative of Oersted's work toward general connectedness and unification, and he also agreed in an emphasis on mediated action. This, however, did not compromise Ampère's basic commitment to inverse-square central forces, and he continued to view these forces as originating in imponderable electrical fluids, positive and negative. Ampère's approach, and the approach of the Continental school that followed him, can be characterized in general terms as a mathematical approach, based on a central-force law, with the latter viewed as expressing the *mediated* interaction of current elements or moving charges. The traditional designation of this school as the "action-at-a-distance" school is thus a bit oblique; one might more appropriately designate it the *charge-interaction school,* as the emphasis is clearly on the interaction of charges according to a mathematical force law, but this is not necessarily a distance interaction.[3] (In what follows, I make use of both the *charge-interaction* and *action-at-a-distance* designations, using the former where precision is necessary, and the latter where common usage will suffice.)

Michael Faraday, like Ampère, took up the imperative toward connectedness and unification generated by Oersted's work. Faraday, however, came out diametrically opposed to the basic elements in Ampère's charge-interaction approach (the electrical fluids, the central-force law, and the sophisticated mathematical technique). Faraday's skepticism concerning the existence of the electrical fluid or fluids was not uncharacteristic for that time – many were questioning the eighteenth-century imponderable fluids. Faraday himself had started out in electricity and magnetism as an autodidact, and his skepticism had developed in that context: The autodidact, confronted with a variety of opinions, and having no teacher to indicate which opinion to choose, can profitably adopt the stance of believing only in the facts, rejecting all hypotheses and theories. In particular, as Faraday wrote to Ampère, the electrical fluids had to be regarded as suspect when viewed from a skeptical standpoint of this sort; to further build conjecture upon hypothesis, by envisioning the fluids as circulating about molecules in magnetized materials, was more than Faraday could swallow:

> I am naturally skeptical in the matter of theories and therefore you must not be angry with me for not admitting the one you have advanced immediately. Its ingenuity and applications are astonishing and exact but I cannot comprehend how the currents are produced and particularly if they be supposed to exist round each atom or particle and I wait for further proofs of their existence before I finally admit them.

Given this rejection of the electrical fluids and the molecular currents, Ampère's explanation of magnetic forces was rendered nugatory: The central-force law that it involved became irrelevant to an understanding of most magnetic phenomena, and the associated sophisticated mathematical discourse was rendered superfluous – a circumstance congenial to Faraday, because he did not have the mathematical background to deal with that aspect of Ampère's theory.[4]

Faraday's alternative to the charge-interaction approach of Ampère was an emphasis on the spaces around the wires and magnets as constituting the ultimate seat of electromagnetic phenomena. According to Faraday, one saw in the spaces surrounding the wires circular patterns of force; these circular patterns were directly and immediately detectable and were, on epistemological grounds, to be regarded as more basic than the push-pull forces that Ampère invoked. Faraday had carried out experiments to demonstrate such circular forces distributed in space – involving magnets revolving about wires, or, alternatively, wires driven in circular paths around magnets – and he was convinced of the primacy of these circular "powers" in space in determining magnetic phenomena.[5] Faraday used the method of iron filings to visualize the patterns of force in space, and later he began to refer to these spatially distributed powers as "lines of force," together constituting a "field" in space.[6] In the full development of Faraday's theory, which included both magnetic and electrical lines of force, electric charges were to be regarded as epiphenomena – manifestations of the termination points of lines of force – having no independent or substantial existence; the electric current, accordingly, was to be viewed not as manifesting the flow of actually existing electrical fluids but rather as constituting an "axis of power," reflecting the dynamics of "a certain condition and relation of electrical forces."[7] In Faraday's approach, then, it was the field that was truly primary, and the approach of the British "field-theory" school following Faraday can more precisely be designated as the "primacy of the field" or *field-primacy* approach.[8] (In what follows, I make use of both the *field-primacy* and *field-theory* designations, once again using the former where precision is called for, and the latter where common usage will suffice.)

As the followers of Ampère and Faraday developed their respective approaches through the rest of the nineteenth century, there were to be variants and nuances within each tradition – without, however, compromising the basic orientation of each. Thus, within the Continental school, the interaction of charges was regarded by some as mediated – by an ether or other agency – and by others as a direct action at a distance. In either case, however, it was the interaction of charge or current elements that was regarded as basic, and the charge-interaction perspective was in this way maintained. Likewise, within the British school, some viewed the field in abstract terms, whereas others viewed the field as embodied in a mechanical ether; both groups, however, continued to view the field as primary, with charges and currents emergent from it, thus maintaining the field-primacy perspective within the British school.[9]

Faraday's discovery of electromagnetic induction in 1831 set a new task for electromagnetic theory, whether of the charge-interaction or the field-primacy variety. Faraday, following up the idea of the connectedness of all with all, had made various attempts in the course of the later 1820s to produce the inverse of Oersted's phenomenon: Whereas Oersted had produced magnetism from electric current, Faraday now believed that one should be able to produce electric currents from magnetism. The key to this, as Faraday discovered in 1831, is that a *changing* rather than a static magnetic field or magnetic situation is required in order to induce an electric current. Including Faraday's new discovery, there were then four basic areas in electricity and magnetism that would have to be treated in an integrated fashion by any general theory: (1) electrostatics – forces between electric charges; (2) magnetostatics – forces among magnetic poles and magnetic materials; (3) electromagnetism – magnetic effects of electric currents; (4) electromagnetic induction – electrical effects of changing magnetic fields or situations. The range of phenomena that would have to be treated by a unified theory was further enlarged in the course of the 1830s and 1840s through a variety of new discoveries, principally by Faraday: His discovery of the effects of dielectric media on relationships between electric charges and electric forces set new tasks for any theory of electrostatics, his discovery of diamagnetic and paramagnetic materials provided new questions for magnetostatics, and his investigation of the various circumstances of electromagnetic induction, as produced by various motions or alterations of the magnets and circuits involved, placed additional demands on any theory of electromagnetic induction. As these new phenomena could be subsumed under the four areas already delineated, the basic task – as perceived by the researchers of the time – remained the treatment in one integrated theory of these four subfields within electricity and magnetism.[10]

Working within the charge-interaction tradition, Wilhelm Weber, in 1846, formulated an integrated theory – treating all four categories of electromagnetic phenomena – that, when evaluated even from within the field-primacy tradition, came across as "so elegant [and] so mathematical"[11] that it could not be ignored. Following Ampère closely, Weber established as the center of his theory an inverse-square force law, specifying the interaction of particles of electrical fluid; Weber followed Ampère also in viewing the interaction specified by the force law not as an unmediated action at a distance but rather as mediated by an ether consisting of a union of the positive and negative electrical fluids. The force law was formulated mathematically as follows:

$$F = \frac{ee'}{r^2}\left(1 - \frac{v^2}{c_W^2} + \frac{2ra}{c_W^2}\right) \tag{1.2}$$

where F is the magnitude of the central force acting between the two particles, e and e' are their electric charges, r is the distance between them, v is their relative velocity, a is their relative acceleration, and c_w is the ratio between electrodynamic and electrostatic units (larger than the modern ratio of units c by a factor $\sqrt{2}$). Of the three terms in the equation, the first accounts for electrostatic forces; the second, involving the relative velocity, yields the phenomena of electromagnetism, and also the phenomena of magnetism per se (the latter by way of the assumption of Ampèrian molecular currents); the third term, involving the relative acceleration, accounts for electromagnetic induction. Together, the various terms add up to a complete and unified account of the phenomena of electricity and magnetism. The formula was open to criticism on physical grounds – the velocity-dependent and acceleration-dependent terms gave rise to much discussion, as we shall see – but the mathematical unification was nevertheless quite impressive. When Maxwell was doing his work in the 1850s and 1860s, Weber's force law stood as representative of what the Continental tradition had to offer, and Maxwell was much impressed – and indeed a bit intimidated – by the elegant unification of electromagnetic phenomena that it offered.[12]

Faraday, who was led to his discovery of electromagnetic induction in the context of the field-primacy approach, naturally looked for a conceptualization of the effect in those terms; he in fact came up with two alternative conceptualizations. First, there was his concept of the "electro-tonic state," a state of the material constituting the circuit (whether a metallic conductor or other conducting medium), whose changes with time were manifested as electric currents; this concept did not bear immediate fruit, although Maxwell was later able to make something of it.[13] Of more immediate importance was Faraday's interpretation of electromagnetic induction in terms of the cutting of magnetic lines of force: A wire moving in a magnetic field would be cutting across the lines of magnetic force, and this would determine the inductive effect. Faraday later found that the current produced, as measured by a galvanometer used ballistically, would be proportional to the number of lines cut.[14] In this way, the idea of magnetic lines of force, originally applied by Faraday to magnetostatics and electromagnetism, was now extended to embrace electromagnetic induction. Faraday further developed the notion of lines of force, in the 1830s and 1840s, in connection with questions of polarized media; he thereby incorporated the effects of material media into his theories of electrostatics and magnetostatics. With this, the idea of lines of force, electric and magnetic, had been extended so as to give a conceptually unified treatment of all of electricity and magnetism. Faraday's unification was, however, only qualitative: There was a paucity of explicit mathematics in his treatment of electricity and magnetism from the field-theoretic point of view, and he himself was not able to remedy the defect (as vis-à-vis Weber's more mathematical theory). It was therefore left for the followers of Faraday – especially William Thomson and Maxwell – to work at the task of mathematizing Faraday's theoretical approach, so as to arrive at a unified mathematical theory within the field-primacy tradition.[15]

The basic task, then, for the British school – from the 1840s on – was to develop a unified mathematical theory, based on the field-primacy approach, that could compete with Weber's theory of 1846 and later modifications of it; this is how Maxwell perceived the task in the mid-1850s, when he began his work on electricity and magnetism. He brought to this task skills, insights, and attitudes that had developed out of his educational experiences as a student at the universities of Edinburgh and Cambridge and as a disciple of William Thomson.

2   *Maxwell's scientific and mathematical background*

Maxwell embarked on his researches in electromagnetic theory early in 1854 – within a month of completing the mathematical tripos at Cambridge and entering "the unholy estate of bachelorhood"; he was then twenty-two years old.[16] Maxwell had first attended the University of Edinburgh, from 1847 to 1850, completing three years of the four-year course. It had been James David Forbes, professor of natural philosophy, who had exerted the greatest influence on Maxwell at Edinburgh; Maxwell had attended Forbes's course in natural philosophy in both 1847–8 and 1848–9. (Repeated attendance was not unusual for a seriously interested student. The material varied somewhat from year to year, and Forbes divided the class into three divisions, Maxwell having been in the middle division in 1847–8 and the first division in 1848–9.) Of significance also was Maxwell's attendance in the logic and metaphysics classes of William Hamilton, as well as his participation during that period in the activities of the Royal Society of Edinburgh. Maxwell went up to Cambridge in 1850; he interacted closely while there with his mathematics coach, William Hopkins, and also with George Stokes, who was then Lucasian Professor of Mathematics. Finally, the rudiments of a relationship of apprenticeship with William Thomson, centering on electromagnetic subjects, had been established even before Maxwell went up to Cambridge; this was continued in a correspondence that was maintained from 1854 until Maxwell's death in 1879. The Edinburgh and Cambridge experiences, together with Thomson's epistolary tutelage, provided the background for Maxwell's choosing, as a central theme of his research agenda, the mathematical consolidation of the field-primacy approach to electromagnetic theory. Basically, what Maxwell was doing, in choosing this topic, was applying the analytical (mathematical) approach he had learned at Cambridge to scientific subject matter whose importance had been emphasized at Edinburgh, in the context of his research apprenticeship under William Thomson.[17]

Although mathematics certainly was not foreign to the Edinburgh experience, and Maxwell had made a good start in pure mathematics and its application to science even before he went up to Cambridge, it was nevertheless the sustained and intensive instruction, practice, and general immersion in mathematics and its application that he experienced at Cambridge – in his preparation for the tripos examination with William Hopkins, in his studies with George Stokes, and in informal interaction with teachers and peers – that provided Maxwell with the sophisticated tools and perspectives needed for the mathematization of Faraday's electromagnetic theory.[18] Mathematics had been cultivated intensively at Cambridge since the scientific revolution and in the eighteenth century had been dominated by the Newtonian form of the calculus. In the nineteenth century, the Continental analysis, stemming from G. W. Leibniz's form of the calculus, had been imported to Cambridge by John Herschel, George Peacock, and Charles Babbage, and the curriculum had subsequently been oriented, by William Whewell, toward a middle-of-the-road position, in which analytical (Continental) methods would be used, but there still would be substantial emphasis on geometrical (Newtonian) representation and illustration; the center of concern was to be "mixed mathematics," in which mathematics would be studied as applied to and embodied in, especially, astronomy, mechanics, and optics – sciences that had matured and become amenable to mathematical treatment in antiquity and that offered the greatest scope for sophisticated application of mathematics in the nineteenth century.[19]

It was the possibility of applying mathematical techniques and demonstrating their power that made astronomy, mechanics, and optics so attractive in the Cambridge setting: "It is only when the student approaches the great theories, as Physical Astronomy and Physical Optics, that he can fully appreciate the real importance and value of pure mathematical science," opined Hopkins; Stokes sustained the emphasis on these core subjects by example.[20] In this context, the newer physical sciences, which had not at that time been mathematized to the same extent as astronomy, mechanics, and optics, were not generally regarded as appropriate subject matter for the Cambridge undergraduate. Some questions relating to heat, electricity, and magnetism had in fact been introduced into the mathematical tripos at Whewell's urging, but these had rapidly fallen into disuse and were explicitly extirpated from the examination just before Maxwell went up to Cambridge; these subjects, apparently, were viewed as belonging to the class of "investigations which are imperfect, or unsatisfactory or involve questions about which mathematicians are not sufficiently agreed." The mixed mathematics curriculum that Maxwell encountered at Cambridge thus provided him with the tools he would need for consolidating Faraday's electromagnetic theory – the continuum mechanics and associated differential equations that he assimilated in this context were in fact just what he needed – but the motivation for applying these to electricity and magnetism was not especially fostered by the Cambridge milieu, with its narrow emphasis on the classical exact sciences.[21] (The Cambridge mixed mathematics was not unlike what is modernly called *applied mathematics* – a discipline that is concerned with the application of mathematics, but in which research is organized in terms of, and is driven by, the possibilities inherent in the equations and the methods, rather than the particular needs of any given applications area. In *mathematical physics,* while the central concern still is with mathematical techniques, research is driven by the needs of physical theory. In *theoretical physics,* the central concern is with the physical principles and their development, rather than the mathematical techniques per se. Along this spectrum, the Cambridge mixed mathematics would appear to have been closer to applied mathematics than to mathematical physics, and quite distant from theoretical physics.[22])

At Edinburgh, on the other hand, the domain of scientific interest was much more broadly construed. Forbes was professor of natural philosophy, and in his lectures on that subject – and also when he wrote up the contents of the course for the eighth edition (1855–60) of the *Encyclopaedia Britannica* – he made it clear that he regarded natural philosophy as a very broad discipline, having relationships with and shading into mathematics, on the one hand, and technology, on the other; there was thus a spectrum from "Mathematics [to] Natural Philosophy [to] Mechanical Arts" that was not to be "formally [i.e., artificially] subdivide[d]." Referring also to the area he covered as "physical science,"[23] Forbes explicitly defined that discipline in quite broad terms: "Physical Science . . . treats of the phenomena of the universe [and] the application of these to the arts and uses of life."[24] The resulting broadly construed discipline was to include a diversity of subdivisions

> treat[ing] . . . of Analytical Mechanics including Physical Astronomy . . . ; of Astronomy as a science of observation; of Mechanics, with reference to the intimate constitution of matter, including Hydrodynamics, Acoustics, and Civil Engineering [with a large section on the steam engine]; of Optics, or Light; of Heat . . . ; and . . . of Electro-magnetism, including ordinary and Voltaic Electricity, Terrestrial Magnetism, and Diamagnetism the discovery of Dr Faraday.[25]

Forbes was an avowedly devoted admirer of Francis Bacon's approach to science, and the diversity of subjects that Forbes dealt with reflects the Baconian emphasis on breadth in scientific investigation; the strong connection with technology that Forbes emphasized was also explicitly Baconian. The "large and comprehensive science of Electro-magnetism" was featured in Forbes's article as itself constituting a Baconian microcosm, with diverse subdivisions and strong technological connections; and Faraday's contributions to electricity and magnetism were given high praise: "Immeasurably the larger part of what we know with regard to the nature and laws of electricity and of its connection with Magnetism, so far as it has been developed since the discovery of Oersted, is due to the genius and perseverance of one man – Michael Faraday."[26]

In the disciplinary matrix specified by Forbes, then, electricity and magnetism – and the work of Faraday in particular – loomed large and would constitute an admirable subject for research; it is here that we find the basic motivation for Maxwell's concern with electricity and magnetism, and his desire to subject it to Cambridge mathematical methods.[27] Beyond this, such a melding of Cambridge methods with subject matter emphasized at the Scottish universities was endorsed by the example of William Thomson.[28] Thomson, seven years Maxwell's senior, was a fellow Scotsman who had been educated first at the University of Glasgow – where the curriculum in natural philosophy had breadth similar to that at Edinburgh – and then at Cambridge; Thomson had then returned to Glasgow as professor of natural philosophy. Beginning in Cambridge, and continuing after his return to Glasgow in 1845, Thomson defined the research program in electricity and magnetism that Maxwell was to take up a decade later and carry through so successfully.[29] Thomson's initial work in this area was in electrostatics, treated from the charge-interaction perspective. Guided by certain analogies between electrostatics as treated by Laplace and Poisson and heat flow as treated by

Joseph Fourier, Thomson had developed, in part on his own and in part on the basis of his reading of works of George Green and Karl Friedrich Gauss, a mathematical approach to electrostatics that emphasized the spatial distribution and geometrical relationships of electrical forces. This approach made use of the language of partial differential equations and potential theory, including the electrostatic potential function and its derivatives, the associated families of orthogonal curves, and the relevant theorems of Gauss and Green. Using this formalism, Thomson was able to show that the Continental charge-interaction electrostatics and Faraday's calculus of lines of force could both be cast in the same mathematical form, so that they were equivalent, at least in mathematical and operational terms.[30]

At that point, in 1845, Thomson did not feel ready to make a firm commitment in physical terms either to the Continental charge-interaction approach or to Faraday's field-primacy approach; indeed, his accomplishment facilitated an agnostic stance, as one could use the new mathematical formalism without having to choose one physical interpretation or the other.[31] What was important, however, for the further development of the field-primacy approach was that Thomson had shown, for the particular case of electrostatics, how Faraday's calculus of lines of force could be mathematized – including the effects of dielectric media. Thomson later extended this kind of treatment to aspects of the magnetic and electromagnetic cases, but when, in the mid-1850s, Thomson became absorbed in other matters, he had not yet managed to put these developments together into a unified whole. The torch had been passed, however, to Maxwell in 1854, and it was Maxwell who took on the task of developing a complete and unified mathematization of Faraday's electromagnetic theory.[32]

Maxwell's experiences as a student at the universities of Edinburgh and Cambridge and as a disciple of Thomson provided not only the background for Maxwell's choice of research area but also the context of commitment to the mechanical worldview within which Maxwell carried out his work on electromagnetic theory; the Cambridge experience was central in this respect. At Cambridge, and at Maxwell's college (Trinity College) in particular, the Newtonian tradition was strong. In mathematics, an exaggerated Newtonianism had become a major hindrance to the Cambridge program in the eighteenth century; in the nineteenth century, reform and counterreform still had left a strong Newtonian residue. Throughout, Newtonian mechanics formed the overarching physical context for the application of mathematics in the mixed mathematics of the tripos. At the elementary level, direct application of the laws of Newtonian mechanics formed the backbone of the tripos curriculum.[33] At the more advanced level, two highly developed bodies of theory based on Newtonian mechanics were taken as paradigmatic of the power of mathematics in treating physical problems, namely, "Physical Astronomy" and "Physical Optics." These had been designated as *"the great theories"* by Hopkins,[34] and Whewell, in his *Philosophy of the Inductive Sciences*, dignified these two theories, and these alone, with full analyses in terms of "inductive table[s]," elucidating the historical and logical structures of these theories. Physical astronomy – namely, the treatment of the dynamics of the solar system in terms of the Newtonian law of gravitation and laws of motion – demonstrated the power of Newtonian mechanics at the macroscopic level. Physical optics – namely, the wave theory of light, founded on the assumption of a mechanical ether – illustrated the power and the promise of Newtonian mechanics as applied at the microscopic level, to the small particles or differential elements of matter, in order to reduce phenomena that were not evidently mechanical at the macroscopic level to the mechanics of their ultimate, microscopic parts.[35]

There were two basic perspectives that Maxwell carried away from this and applied to his researches on electricity and magnetism. First, and more generally – based on the conjunction of physical astronomy and physical optics, the macroscopic and the microscopic – there was the program of reducing all of physical science to mechanics. Maxwell articulated this program clearly in his inaugural lecture at the University of Aberdeen, as he embarked on his academic career: He espoused an ontology of matter and motion in which "all matter [is] in itself . . . the same, and can be modified only by differences of arrangement and motion and by being actuated by different systems of force." Given this ontology, it followed that "the Electrical and Magnetic sciences," and indeed all of "the Physical Sciences," "treat . . . of . . . phenomena . . . depending on conditions of matter," namely, the aforementioned conditions of "arrangement and motion and . . . being actuated by different systems of force." The aim of the physical sciences, therefore, was mechanical reduction: "As soon as any . . . phenomenon has been reduced to a change in the arrangement and motion of a material system . . . the phenomenon will be physically explained." The goal of Maxwell's work in electricity and magnetism was thus mechanical reduction and explanation, and the work was in fact to be carried out entirely within a mechanical framework.[36] A second, more particular perspective that Maxwell brought away with him from the Cambridge experience reflected the strong emphasis there on the wave theory of light: In the Cambridge setting, one took for granted the existence of a pervasive mechanical medium in space, namely, the luminiferous ether, whose primary function was to support luminiferous undulations – light waves. As will be discussed more fully in Chapter 5, however, it was theorized early on that the ether might have more functions than just the transmission of light, and that it could thus serve as a more general focus of mechanical reduction; Maxwell's later treatment of electromagnetic phenomena in terms of a pervasive mechanical medium – ultimately identified with the luminiferous medium – can thus be seen as a direct sequel of the Cambridge emphasis on the wave theory of light.[37]

The mechanical worldview was not the exclusive preserve of Cambridge University in the nineteenth century, and Maxwell's experiences as a student at Edinburgh and as a disciple of Thomson served to reinforce the Cambridge emphasis on mechanics. There were, however, important differences between the Scottish, Cantabrigian, and Thomsonian approaches to mechanical reduction and mechanical representation, and these reflected differing methodological commitments; the methodological richness of Maxwell's work can be understood against this background.

3   *The methodological background*

Differing attitudes toward mechanical reduction and mechanical representation were conveyed by Maxwell's experiences at Edinburgh, at Cambridge, and with William Thomson. To characterize these in brief: At Cambridge, mechanical representation was intended – at least this was the hope – to give an account of the true, hidden natures of things; at Edinburgh, mechanical representation was viewed more skeptically, as primarily heuristic and analogical; and in Thomson's hands, mechanical representation had a variable role – heuristic and analogical in the early stages of an investigation, and more realistically intended later on. Maxwell in his own work was to make good use of these various approaches to mechanical representation, in response to the varying exigencies of his research program.

The Cambridge approach to the utilization of mechanical representation in physical theory was conditioned by the strong mathematical orientation of the Cambridge program. Mathematical arguments of any complexity necessarily involve long deductive chains, and when such arguments are applied in scientific contexts, the associated deductive chains will be interposed between the basic assumptions of the mathematical theory and the deductive entailments that are to be compared with observation. What results is a type of scientific theory in which the primitive theoretical assumptions, or basic principles, can be quite distant – in a logical and reflective sense – from the empirical data they are intended, ultimately, to address. This kind of scientific theory may be denoted, in a metaphorical sense, *deep theory:* The primitive assumptions lie deep below the empirical surface, distant from the data to which they pertain.[38] This mathematically based inclination toward deep theory was coupled at Cambridge with Newtonian loyalism, which carried with it endorsement of Newton's plan for application of the laws of mechanics to the microcosm, thus building upon his successful celestial mechanics a program for understanding all of the properties, structures, and forces of matter in mechanical terms: "I wish we could derive the rest of the phenomena of Nature by the same kind of reasoning from mechanical principles, for I am induced by many reasons to suspect that they may all depend upon certain forces by which the particles of bodies . . . are either mutually impelled towards . . . or . . . repelled . . . from one another."[39] Theories involving the "particles of bodies" and the forces they exert will necessarily be deep theories in the sense already alluded to, for the particles and forces postulated are distant from the observable macroscopic bodies they are presumed to constitute – distant both in terms of physical scale and in terms of the logical, deductive distance that intervenes between the properties assumed for the particles and the characteristics deduced for macroscopic bodies. Thus, Cambridge mathematicism and Cambridge Newtonianism together militated for an emphasis on deep theory.[40] Finally, the efficacy of this kind of deep theory was conclusively demonstrated – at least to the satisfaction of the Cambridge establishment – by the grand successes of "Physical Astronomy and Physical Optics,"[41] the two highly developed, deep theories that were taken as paradigmatic at Cambridge. The former, based on the law of universal gravitation, was the classical example, universally accepted; the latter, based on the undulatory hypothesis, was the contemporary example, currently new and exciting, and locally championed.

Two Cambridge lights emerged as philosophical spokesmen for this kind of deep theory: John Herschel and William Whewell.[42] Maxwell was to become quite familiar with the opinions of both, and had positive responses to both. Herschel, in his influential *Preliminary Discourse on the Study of Natural Philosophy* (1830), defined the need for deep theory in terms of the quest to uncover the "hidden processes of nature":

> The immediate object we propose to ourselves in physical theories is the analysis of phenomena, and the knowledge of the hidden processes of nature in their production, so far as they

        can be traced by us. An important part of this knowledge con-
        sists in a discovery of the actual structure or mechanism of the
        universe and its parts.

For Herschel, then, mechanical representations or models were to be
devised not as mere analogies or heuristics, but as a part of the quest for
knowledge of the "actual structure or mechanism of the universe." The
quest would not be an easy one:

        Now, the mechanism of nature is for the most part either on
        too large or too small a scale to be immediately cognizable by
        our senses; and her agents [forces] in like manner elude direct
        observation, and become known to us only by their effects.

How, then, to proceed?[43]

The process that Herschel recommended was hypothetico-deductive in
structure, beginning with the "framing [of] hypotheses," and proceeding
thence to the "constructi[on of] theories." A "well imagined hypothesis"
would be the starting point, and it would be well if the hypothesis were in
some sense or to some extent inductively based, but it would be permissi-
ble also simply to "form . . . a bold hypothesis," and if the theory turned
out to be successful, it would "matter . . . little" how "its postulates . . . should have been fixed upon," or "how it ha[d] been originally
framed." Herschel, then, urged some caution, but in fact allowed considerable latitude in the initial choice of hypotheses; indeed, even hypotheses that appeared "complex," "artificial," or "strange . . . at first sight"
might be entertained. The proof of the pudding, for Herschel, was in the
eating. After framing the hypotheses and drawing the deductive consequences of these, one would finally compare with observation in order to
verify or falsify the hypotheses and the theories based upon them: The
"suppositions" of a theory are "verified by the coincidence of the conclusions we shall deduce from them, with facts"; again, we "form . . . a
. . . hypothesis . . . and try . . . the truth of it by following out its consequences and comparing them with facts." If the "postulates" of a theory, even though they are "strange . . . at first sight," nevertheless "lead
us, by legitimate reasonings to conclusions in exact accordance with
numerous observations . . . made under . . . a variety of circumstances
. . . we cannot refuse to admit them." If, further, a theory not only were
to include among its entailments the known facts but also were able to
"*predict facts before trial*," this would constitute verification of an especially convincing kind. In any case, it was this kind of hypothetico-
deductive approach that would enable the Cambridge scientist to construct theories dealing with the "hidden processes of nature" and discover

the "actual structure or mechanism of the universe and its parts," without ever gaining direct access to the "secret recesses" in which these hidden processes "are effected."[44]

In order to have a measure of confidence in a theory established in this manner, Herschel felt that it would be necessary for the theory to "truly represent *all* the facts, and include *all* the laws, to which observation and induction lead." Most theories, he conceded, were not quite so comprehensive in their coverage, and one could not make credible reality claims concerning the hypotheses on which such theories rested; in such cases, the hypotheses might well have heuristic value, but one would not want to become overcommitted to them:

> Regarded in this light [as heuristics], hypotheses have often an eminent use: and a facility in framing them, if attended with an equal facility in laying them aside when they have served their turn, is one of the most valuable qualities a philosopher can possess; while, on the other hand, a bigoted adherence to them, or indeed to peculiar views of any kind, in opposition to the tenor of facts as they arise, is the bane of all philosophy.

There were, however, some "limited instances" in which it had been possible to develop a higher degree of confidence in a hypothesis:

> It may happen (and it has happened in the case of the undulatory doctrine of light) that such a weight of [evidence] may become accumulated on the side of an hypothesis, that we are compelled to admit one of two things; either that it is an actual statement of what really passes in nature, or that the reality whatever it be, must run so close a parallel with it, as to admit of some mode of expression common to both.

Thus, to within semantic equivalence, one could, in the case of a well-developed and well-verified theory such as the undulatory theory of light, arrive at the judgment that the fundamental hypotheses of the theory describe "what really passes in nature."[45]

Herschel, then, admitted that there were dangers in a hypothetical approach; the payoff of the hypothetical approach in favorable cases, however, would be true insight into the "reality" of the "hidden processes" and "mechanism[s] of nature." Herschel's central example of such a favorable case was the undulatory theory of light, and, indeed, Herschel's methodological views had developed in conjunction with his growing acceptance of, and confidence in, the wave theory. Others at Cambridge experienced similar conversions to the wave theory of light in the 1820s and 1830s and drew similar methodological conclusions: Prominent among these were William Whewell, George Airy, James Challis, and, a few years later, George Stokes. Together, this group established and expressed the Cambridge attitude toward the wave theory of light and the methodology of mechanical representation in general.[46] Whewell wrote extensively on the subject of methodology, and although there was much in Whewell's approach that overlapped with Herschel's hypo-thetico-deductive model of scientific reasoning,[47] Whewell added an important intuitionistic element in his own account of scientific methodology. For Whewell, the principles of Newtonian mechanics and the ontology of matter and motion had a priori status, and there was, by the nineteenth century, no remaining element of tentativeness in them. For Whewell, then, even more than for Herschel, it was possible to arrive at certain truth concerning the ultimate mechanical basis of natural phenomena.[48]

Particularly characteristic of Whewell's approach was the notion of "consilience of inductions," in which two sets of phenomena, not previously seen as being related, would be united in terms of one set of basic ideas – one theory; both the intuitive appeal and the pragmatic power of the resulting unified theory would then tend to validate that theory. Thus, for example, a connection between electromagnetism and light – the possibility of which Whewell himself considered – would be such a consilience and would tend strongly to validate the theory that achieved it.[49] For Whewell, then, broad, comprehensive, deep theory, encompassing many areas, was essentially self-justifying: If this kind of grand unification could be achieved, the resulting comprehensive theory would be seen – albeit not in a flash of insight, but rather in an "intuition [that] is progressive" – to be "not only consistent with the facts, but necessary."[50]

Maxwell was familiar with and enthusiastic about the ideas of both Herschel and Whewell. Maxwell probably had been introduced to the ideas and writings of these philosophers even before he arrived in Cambridge. At Cambridge, Whewell was master of Maxwell's college (Trinity), and Forbes had written to Whewell, recommending Maxwell to him as worthy of attention; explicit and implicit references to Whewell, dating from late in Maxwell's Cambridge period, attest to his familiarity with and approval of Whewell's ideas at that time.[51] Maxwell probably had already been introduced to Herschel's *Preliminary Discourse* at Edinburgh, for we know that the book was regularly used in Forbes's natural philosophy class: "I have read Herschel's collected Essays, which I like much," Maxwell reported in 1858, and Herschel's work in optics and other areas was of relevance to, and was noted by, Maxwell.[52] Beyond that, Maxwell had been immersed in the Cambridge scientific culture that Herschel and Whewell – as well as Stokes, Airy, Challis, and others – had both formed and articulated. Their methodological views, therefore, were prominent among the methodological alternatives on which Maxwell could draw. There was, however, an alternative methodological heritage to which Maxwell was exposed: a variety of Scottish skepticism, promulgated by Forbes at Edinburgh, that was in strong contrast to the Whewellian intuitive certainty and the general Cantabrigian confidence in deep theory.

Both Herschel at Cambridge and Forbes at Edinburgh invoked the name of Francis Bacon in establishing the genealogies of their respective approaches, but there was a great difference in how they actually used Baconian methods. Herschel did pay lip service to the inductive ladder that was the foundation of the Baconian approach: "There can be no doubt . . . that the safest course, when it can be followed, is to rise by inductions . . . from law to law." As concerned the most interesting questions, however, dealing with the "hidden processes of nature," that "safest course" could not be followed, and Herschel, as we have seen, recommended instead a hypothetico-deductive approach that was quite different from the inductive ladder, involving hypothetical leaps of the type that Bacon had explicitly condemned.[53] Forbes, however, was an honest Baconian, in a variety of ways. To begin with, Forbes, as we have already seen, construed the discipline that he practiced and taught in the broadest possible manner, embracing all of physical science, along with related mathematical ideas, on the one hand, and technological concerns, on the other. With respect to this vast array of subject matter, the deep theories favored at Cambridge stood feeble and impotent at the midpoint of the nineteenth century. If one selected subject matter on the basis of its amenability to treatment by Cambridge methods – as the Cambridge scientists did – those theories would be found to serve admirably as organizing principles. If, however, one was concerned with the whole range of physical science, a theoretical and deductive arrangement of the subject matter would be beyond reach, and one would have to settle instead for an encyclopedic arrangement; Forbes's account of the physical sciences was of this latter kind – much closer to the kind of "natural and experimental history" that Bacon had favored than to the kind of deductive treatment presented in the *Principia* of Newton (or Descartes) and favored at Cambridge.[54]

Beyond the constraint on theoretical organization imposed by the goal of broad coverage, Forbes in certain cases made active choices in the direction of declining to make use of existing deep theory, maintaining instead a skeptical stance. For example, Forbes presented electricity and magnetism in encyclopedic fashion, with no suggestion that there was any unifying theory that could be fruitfully applied. The Continental unification failed at the outset for Forbes, for he found the complete reduction of magnetism to electricity through the hypothesis of molecular currents to be "arbitrary and improbable" and unlikely to constitute a "true and complete physical picture of the condition of magnetized bodies." Furthermore, Forbes believed that diamagnetism was not convincingly explicable on the Ampèrian hypothesis. As for the followers of Ampère – Franz Neumann, Wilhelm Weber, and others – Forbes had not noticed their work: "Little or nothing has been added either to the theory or to the deductions from it since [Ampère's] death."[55] Faraday's work did receive detailed attention from Forbes (taking up six pages of the *Britannica* article), but there was little mention of the concept of the line of force as a unifying theme. Instead, Faraday's studies of electricity and magnetism were depicted as constituting a Baconian adventure of the most admirable kind – unmotivated by theoretical considerations. Faraday's work on electromagnetic induction represented a "perfect specimen of inductive sequences"; his work on diamagnetism demonstrated "an unrivalled skill in the application and invention of experimental methods." One aspect of the line-of-force concept was indeed mentioned by Forbes, but only for the purpose of dismissing it: "The part of Dr Faraday's conclusions, however, most open to exception, is what refers to electric action at a distance, which he conceives to depend *solely* upon induction acting on intervening particles . . . along curved lines." Skeptical concerning the fundaments of Faraday's theoretical approach, Forbes wrote to Maxwell in 1857 that he was "by no means yet a convert to the views which Faraday maintains." Forbes was thus not prepared to make use of either the Continental option or Faraday's approach as a basis for the unification of electromagnetic theory; he was skeptical of both, seeing adherence to either as contrary to Baconian scruple.[56]

Examples of Forbes's antipathy toward deep theory and his associated commitment to an empirical and encyclopedic approach in science can be multiplied. The principle of conservation of energy and the associated kinetic theory of heat, which were widely perceived as major integrating themes in the physics of the 1850s, were not used by Forbes to integrate his treatment of heat. He regarded the claims made for these as at least somewhat premature: More work was "still required to justify all the conclusions which the zealous promulgators of this comparatively new 'mechanical theory of heat' have advanced." As for the undulatory theory of light, although Forbes assented to the basic wave hypothesis, he purveyed with relish tales of the disagreements among the various investigators attempting a mathematical ether dynamics: "One theorist assumes that the density of ether is the same in all bodies, another that it is greatest in a vacuum, and others precisely the reverse; one that vibration is not accompanied by change of density, another that it is; and so on in almost endless variety." Continuing in that vein, Forbes judged that "a great part of this vast mathematical toil [on ether dynamics] has been without immediate result in optics. It is by comparing the conclusions arrived at by authorities of seemingly equal weight, that we learn the difference between a stable physical induction and a clever mathematical hypothesis." Thus, what were seen by others as major organizing theories of midcentury physical science provided for Forbes occasion for skeptical and derogatory remarks.[57]

Forbes himself explicitly traced this approach to science back to Bacon: "Whatever has been learnt or discussed concerning the means of arriving at truth in Natural Science, it is not pretended that we have recently become possessed of any canons or rules of discovery superseding those fundamental principles of observation and experiment so well laid down by Bacon." In recent historical scholarship, the Scottish approach to science, including aspects of the skeptical attitude displayed by Forbes, has been traced also to the Scottish tradition in philosophy, going back to David Hume, extending through the Common Sense school, and reaching ultimately to Sir William Hamilton, whose moral philosophy class at Edinburgh Maxwell attended. In any case, this kind of skepticism was seen as peculiarly Scottish by those involved; William Thomson expressed his skepticism concerning a certain theory by delivering what he chose to describe as "the Scottish verdict of *not proven*." Along with the skeptical element, however, the Scottish philosophical tradition also provided a positive element, in the form of an emphasis on analogies. Favored by the relational emphasis of Common Sense epistemology, analogies were an important part of Forbes's armamentarium of conceptual apparatus to be deployed in ordering scientific data, filling in for the hypothetical and theoretical reasoning that he rejected, while involving less commitment in the way of ontological claims than would a hypothetical approach.[58]

Paradigmatic for Forbes was an analogy between radiant heat and light, which had been established principally through the work of Macedonio Melloni in the early 1830s. Melloni had carried out experiments on the reflection, refraction, and absorption of radiant heat and had concluded that there exist different varieties of radiant heat, analogous to the different colors of visible light. Forbes himself followed up the analogy further, demonstrating experimentally, in 1836, that radiant heat displayed polarization properties analogous to those displayed by light. Forbes commented at that time that "the importance of analogies in science has not perhaps been sufficiently insisted on by writers on methods of philosophizing."[59] Forbes was also able to make good use of the analogical approach in his work on glaciers, where he compared the motion of a glacier with the flow of a fluid. In addition, Forbes was concerned with the analogies between magnetism and electricity; he remarked on the "analogies of heat, light, and electricity"; and he considered Faraday's discovery of electromagnetic induction to be the result of attention to the analogies suggested by Oersted's discovery.[60]

Forbes's use of analogy represented, to a significant extent, a reaction against and a flight from the kind of ontologically committed, hypothetico-deductive, mathematical theorizing practiced at Cambridge. There was, however, another way in which the analogical approach could be used, not as an exclusive alternative to the Cambridge methodology but as an adjunct to it. It was William Thomson who pointed out this middle road to Maxwell, by example as well as by precept. Thomson himself benefited from combined Scottish and Cambridge backgrounds, and he was thus well prepared to make use of a combination of Scottish and Cambridge methodologies. Thomson's approach, however, represented something more than a mere concatenation of Scottish and Cambridge influences: Thomson took the mechanical theory of heat very seriously – which neither Forbes nor the Cambridge group did – and that gave a particular cast to Thomson's use of mechanical representation.[61] Thomson's synthetic approach, in turn, had definite implications for Maxwell, for when Maxwell set out on his own studies of electromagnetic theory, he was most directly following Thomson's example.

As discussed in Section 2, the initial thrust of Thomson's use of analogy in electricity and magnetism had been essentially mathematical: In 1842, his aim had been to establish a mathematical analogy between heat flow and electrostatics, in order to be able to carry over some mathematical results from the thermal case to the electrical case; the analogy was thus entirely heuristic and not intended to establish any theory of electricity. In 1845, Thomson used this analogy to demonstrate a mathematical equivalence between Faraday's lines-of-force formalism and the traditional charge-interaction electrostatics; again, the basic thrust was agnostic, in showing that no choice between the two theories needed to be made or could be made on the basis of mathematical agreement with empirical results. Beginning in 1845, however, Thomson allowed himself some speculations about the possible theoretical implications of what had started out as a mere analogy. If one took the mathematical analogy between heat flow and electrostatics to be suggestive of a physical, on-tological similarity, the implication would be that just as heat flow de-pends on an "intervening medium," so do electrical forces:

> It is, no doubt, possible that [electrical forces] may be discov-ered to be produced entirely by the action of contiguous parti-cles of some intervening medium, and we have an analogy for this in the case of heat, where certain effects which follow the same laws are undoubtedly propagated from particle to parti-cle.[62]

Thus, for Thomson, the use of analogy was not the permanent strategy of a perennial skeptic, but rather possibly a step along the way toward a characterization of the actual physical situation; that turn of Thomson's thought was to be paradigmatic for Maxwell.

Furthermore, beginning in 1845 and developing more fully in 1847, Thomson's formalism for electrostatics incorporated the related concepts of mechanical effect and *vis viva*. In a pioneering application of the nascent energy formalism, Thomson in 1845 began working with a func-tion that expressed the total energy or "mechanical effect" of an elec-trostatic system: He showed that forces could be derived in terms of derivatives of this function with respect to spatial displacements of the electrified bodies; he showed that this function could be expressed as the integral of the square of the force field over the system; and he showed how all of this could be interpreted by reference to mechanically analo-gous systems in which fluid flows represented the lines of force, and the energies or mechanical effects involved were exhibited as the kinetic energies or *vires vivae* of the fluids.[63] As the energy idea matured in the later 1840s and came to dominate the scientific agenda of the 1850s, Thomson and then Maxwell came to see the energy connection as furnish-ing one of the prime motivations for constructing mechanical representa-tions of electromagnetic phenomena and for regarding these as having realistic status, with the energies of the mechanical media posited taken to constitute the actual energies of the electromagnetic systems under con-sideration. Throughout his career, Maxwell was to regard these argu-ments from energy considerations as paramount.[64]

Maxwell was thus prepared – by his experiences at Edinburgh, at

Cambridge, and with William Thomson – to bring a diversity of methodological approaches to bear in his attempt to unify electromagnetic theory from a field-theoretic point of view. These methodological options ranged from the analogical approach recommended by Forbes's skepticism and Scottish Common Sense philosophy, through the middle way represented by the synthetic approach of William Thomson, to the fully committed search for the hidden mechanisms of nature as championed by Cambridge philosophers such as John Herschel and William Whewell.

2

Mechanical image and reality in Maxwell's electromagnetic theory

As he pursued the task of constructing a unified account of electromagnetic phenomena from a field-theoretic point of view – from his initial explorations under Thomson's tutelage in 1854 through his work on a second edition of the *Treatise on Electricity and Magnetism* in the months before his death in 1879 – Maxwell was unwavering in his basic commitment to a broad mechanical framework, within the confines of which this task was to be carried out. Within this broad mechanical framework, however, there were various methodological options at Maxwell's disposal – traceable to his experiences at both Edinburgh and Cambridge, and also to his interaction with William Thomson – and Maxwell was to make full use of this variety of options, in response to the shifting needs of his evolving research program. In brief, Maxwell started out using an analogical approach to mechanical representation, rooted in Scottish skepticism and reflecting a desire to proceed with minimal physical commitment at the outset; in this context, he presented the mechanical images in his first major paper on electromagnetic theory, "On Faraday's Lines of Force" (1855–6), as purely illustrative, with no claim whatever to realistic status. Subsequently, responding to William Thomson's judgment that the time had come to go beyond mere analogy in electromagnetic theory and to begin the task of constructing a realistic mechanical theory, Maxwell developed his molecular-vortex representation of the electromagnetic field in the paper entitled "On Physical Lines of Force" (1861–2). In that paper he made an explicit commitment to the probable reality of the basic features of the molecular-vortex hypothesis in a manner characteristic of the Cambridge school. Finally, in the later 1860s and the 1870s, Maxwell began a measured retreat from his realistic commitment to the molecular-vortex model, without ever completely giving it up; his attitude toward mechanical representation in that period was complex and nuanced – not reducible to one or the other of the initial options. An understanding of these successive phases in Maxwell's utilization of mechanical models, as well as the varying positions that he adopted with respect to the question of the reality of these mechanical images, is important for our general understanding of the role of mechanical models in nineteenth-century science and also provides the broad context for an understanding of Maxwell's stance with respect to the molecular-vortex model and his utilization of it in connection with his major innovations in electromagnetic theory.

1   *Maxwell and the uses of analogy*

In his paper "On Faraday's Lines of Force" – his first major effort in electromagnetic theory – Maxwell made use of mechanical representation in an analogical sense, at a time when he was not yet ready for a deeper commitment to any mechanical picture. In part, Maxwell supported his choice of an analogical approach, with its avoidance of ontological commitment, by invoking the kind of Scottish skepticism promulgated by Forbes at Edinburgh.[1] For Maxwell, however, the analogical approach was to be employed not as a permanent alternative to the kind of ontologically committed deep theory favored at Cambridge but rather as a prelude to that kind of theorizing; in this, Maxwell followed the example of Thomson rather than Forbes. Maxwell found Thomson's way of using analogies especially attractive, as he indicated in a letter to Thomson: "Have you patented that notion with all its applications?" asked Maxwell in May of 1855, "for I intend to borrow it for a season."[2] Borrow it he did, along with the notion that its primary value was as a temporary expedient, as a prelude to something better.

Maxwell saw himself in a situation in which he could make good use of just such a temporary expedient. "The present state of electrical science," he felt, was "peculiarly unfavourable to speculation." Much was known about electricity, but that knowledge was scattered and fragmentary, and Maxwell could not yet see his way through to a theoretical structure, based on the field approach, that would unify all the ramified phenomena of electricity and magnetism. One might be tempted, in such a situation, to "adopt a [working] hypothesis," which might at least lead to a "partial explanation" of electromagnetic phenomena. In Maxwell's opinion, however, that would be dangerous, because it might foster premature commitment: "If [in this situation] we adopt a physical hypothesis, we see the phenomena only through a medium, and are liable to that blindness to facts and rashness in assumption which a partial explanation encourages." Evincing here more scruple with respect to hypotheses than Herschel or

Whewell – who allowed provisional hypothesizing and theorizing, if carried out with appropriate care – Maxwell showed himself more the follower of Forbes in his concern to "avoid the dangers arising from a premature theory," to avoid being "carried beyond the truth by a favourite hypothesis." In this situation, the analogical approach seemed particularly well suited as a tool of investigation, for it would permit the use of a mechanical representation, which would aid in the task of "simplification and reduction of the results of previous investigation to a form in which the mind can grasp them," but "without being committed" in any way to the literal truth of that mechanical representation. (The possibility of proceeding without any mechanical representation, in a purely mathematical vein, was rejected by Maxwell out of hand, as he believed that disembodied mathematics was bound to be unfruitful; Scottish, Cambridge, and Thomsonian traditions were in agreement on that point.[3])

Merely as an analogy, then, Maxwell proposed a mechanical representation in which an incompressible fluid was pictured as flowing through a porous medium. In "Faraday's Lines," this image was applied to the elucidation of electric fields, magnetic fields, and distributions of electric current; the flow lines of the incompressible fluid were taken to correspond to magnetic lines of force, electric lines of force, or lines of electric current, depending on the particular problem being analyzed. The flow analogy could provide no image of the coexistence and interaction of electric fields, magnetic fields, and electric currents; rather, the flow analogy provided segmented, compartmentalized understanding of each of the three electromagnetic phenomena, each considered in isolation from the other. Such compartmentalization would have been intolerable from the point of view of providing a theory: "No electrical theory can now be put forth, unless it shews the connexion not only between electricity at rest and current electricity, but between the attractions and inductive effects of electricity in both states." The flow representation, however, did not have to be measured against that high standard, for it was intended as a mere analogy, with no claim to either comprehensiveness or truth value. Maxwell explicitly and repeatedly cautioned the reader that the incompressible fluid referred to was an "imaginary fluid," and "not even a hypothetical fluid." The mathematical isomorphism between the equations of percolative streamline flow and the equations describing electric or magnetic lines of force was nothing more than that, and no "physical theory," no specification of the actual "physical nature of electricity" or magnetism, was implied.[4]

There were further reasons for Maxwell's modest stance Wilhelm

Weber's "professedly physical theory of electro-dynamics," formulated in the action-at-a-distance (or charge-interaction) tradition, was by Maxwell's own admission "so elegant [and] so mathematical" that it could not be ignored. Weber had been able to develop a theory that gave a coherent and connected account of the basic phenomena of electricity, electromagnetism, magnetism, and electromagnetic induction – basically all of the phenomena of electricity and magnetism. Maxwell was not able at that point to propose an alternative theory of comparable range based on the field approach, and he evidently was concerned that strong claims in favor of a partial theory of his own – as against Weber's elegant, mathematical, and very comprehensive theory – would invite criticism and perhaps ridicule. Maxwell therefore presented his mechanical picture merely as an analogy – a heuristic device and "temporary" expedient – justifying his effort on grounds of theoretical pluralism: "It is a good thing to have two ways of looking at a subject, and to admit that there *are* two ways of looking at it."[5]

There was other ground on which Maxwell believed he might be criticized: He was presuming to contribute to the theory of electricity and magnetism, while having made no contribution whatever to the experimental side of that field. Only in the twentieth century has the theoretical physicist per se had an acknowledged role; given nineteenth-century norms, Maxwell felt constrained to be both apologetic and moderate in his claims: "By the [analogical] method I adopt," Maxwell wrote, "I hope to render it evident that I am not attempting to establish any physical theory of a science in which I have hardly made a single experiment." Maxwell's deference, one surmises, was primarily toward Faraday, who was eponymously honored in the title of Maxwell's paper; furthermore, Maxwell's main competitor in the realm of mathematical theory – Wilhelm Weber – added to his theoretical prowess impressive experimental credentials. Having no such credentials as an electrical experimenter himself, Maxwell did not presume to present a "true solution" to the problems of electrical science; instead, he offered a heuristic analogy, defining for himself an auxiliary role vis-à-vis the experimental philosophers, helping but not usurping:

> If the results of mere speculation which I have collected are found to be of any use to experimental philosophers, in arranging and interpreting their results, they will have served their purpose, and a mature theory, in which physical facts will be physically explained, will be formed by those who by interrogating Nature herself can obtain the only true solution of the questions which the mathematical theory suggests.[6]

This statement expressed not only the modesty of Maxwell's aims in "Faraday's Lines" but also the hopes he had for the future. He looked forward to something better than mere analogy: He looked forward to "a mature theory, in which physical facts will be physically explained." Thus, his aim in "Faraday's Lines" had "not [been] to establish [a] theory," but his hope for the future was that "a . . . theory . . . will be formed"; the fluid-flow analogy had "not [been] introduced to explain actual phenomena," but the hoped-for theory was to be one "in which physical facts will be physically explained."[7] In 1855, Maxwell was a young man without reputation, diffident and deferent with respect to both Faraday and Weber; he was just beginning to make headway in the task of understanding electromagnetic phenomena and wanted to avoid premature commitment; and he found himself able to devise only a segmented and compartmentalized mechanical representation of electromagnetic phenomena. In these circumstances he made no strong claims for the flow representation, putting it forward merely as an illustrative and heuristic analogy; he looked forward, however, to a better time.[8]

2   *Toward a realistic, comprehensive, and explanatory theory*
    The signal that the time had come to go beyond mere analogy and begin to talk in earnest about the nature of things came from William Thomson. Thomson had pioneered the use of physical analogies in electromagnetic theory, and Maxwell had followed him; Thomson had, from the outset, regarded these analogies as merely preliminary steps along the way toward a hoped-for "physical theory,"[9] and Maxwell had agreed; finally, in 1856, Thomson decided that the time had come to talk about the nature of things in electromagnetic theory, and Maxwell was to follow him once again. Maxwell was no blind follower – we have seen that he had his own good reasons for his use of the method of analogy – but he was nonetheless a devoted follower of Thomson, and it is not surprising to find Maxwell following his mentor through this whole sequence.
    It was within a year of the publication of Maxwell's "Faraday's Lines" that Thomson decided the time had come to go beyond mere heuristic analogy in the mechanical representation of electromagnetic phenomena. In a paper published in 1856 he announced that he was ready to propose a description of "reality," of the "ultimate nature of magnetism."[10] This new departure must be understood against the broader background of Thomson's developing commitment to a "dynamical" understanding of physical phenomena. That commitment had roots in Thomson's earlier discussions of "mechanical effect" in electrostatic systems, but it began to assume a central position in his thinking only as a result of his encounter

with James Prescott Joule at the British Association meeting of 1847, where he "learned from Joule the dynamical theory of heat, and was forced to abandon at once many, and gradually from year to year all other, statical preconceptions regarding the ultimate causes of apparently statical phenomena."[11] Thomson's conversion to the dynamical theory of heat was in fact not quite so abrupt; he continued to defend the caloric theory against Joule's novel views for some years after 1847. In essence, however, Thomson's recollection was correct: He did experience a dramatic conversion to the dynamical theory of heat – by 1851 if not in 1847 – and the general lesson that he abstracted from that conversion experience was that all phenomena are ultimately "dynamical."[12] For Thomson, a "dynamical" theory, as opposed to a "statical" theory, was one in which the forces – and hence effects in general – exerted by a given physical system were referred to internal motions within that system, rather than to primitive attractions or repulsions between its particles. Thus, in Thomson's paradigmatic case of the dynamical theory of heat and gases, gas pressure was the result of internal motions, rather than static repulsive forces between caloric particles within the gas.[13] Thomson, in his dynamical theory of heat, followed Humphrey Davy and W. J. M. Rankine in assuming that the motions that constitute heat are rotary motions associated with individual molecules – "molecular vortices," in Rankine's terminology. Surrounding each "molecular nucle[us]," then, there is vortical motion of the material medium that "interpermeat[es] the spaces between molecular nuclei." The nature of this material medium was not precisely specified; it might be, for example, a "continuous fluid," or it might be a molecular fluid.[14]

The second major input to Thomson's growing conviction that he was now able to specify the "ultimate nature of magnetism" went back to Faraday's discovery, in 1845, of a magnetic action on light. Faraday had discovered that a beam of light propagating in a piece of glass situated in a magnetic field would experience a rotation of its plane of polarization.[15] What was particularly striking to Thomson was that the handedness of the rotation depended on the direction of propagation of the light ray: If a beam of light propagating in one direction through the magnetic field were to experience a right-handed rotation of its plane of polarization, then a beam propagating in the opposite direction would experience a left-handed rotation. The power of optical rotation previously known for certain media, such as sugar solutions, had a definite handedness – dextrorotary or levorotary, depending on the isomer – independent of the direction of propagation of the light beam; such optical rotation could be explained by assuming that the dissolved material consisted of tiny "spiral fibers," of definite handedness, which would rotate the plane of polarization with that given handedness. The Faraday rotation, however, could not be explained in this way, because its handedness was variable. Thomson argued that this could be explained only on the assumption that the magnetic line of force corresponds to an axis of rotation in the medium through which the light propagates: A definite sense of rotation in space would give opposite handedness when referred to opposite directions of propagation. This, in turn, provided for a connection with Rankine's theory of molecular vortices: Thomson concluded that the actual mechanical condition characterizing a region traversed by magnetic lines of force would be one in which the axes of the molecular vortices would all be aligned in one direction, that being the direction of the line of force.[16] A few years later, in a talk at the Royal Institution, Thomson asserted that "a certain alignment of axes of revolution in this [vortical] motion IS *magnetism*. Faraday's magneto-optic experiment makes this not a hypothesis, but a demonstrated conclusion."[17]

The payoff for that way of looking at magnetism, Thomson hoped, would be a unified and realistic theory of electromagnetic phenomena, developed from the field-theoretic point of view, and mechanically founded on the image of molecular vortices. Magnetic forces would be explained not as the result of the static interaction at a distance of magnetic poles or electric currents but rather dynamically, as a result of vortical motions in the intervening medium – the dynamical approach here allying with the field-theoretic outlook in a powerful and mutually reinforcing combination. Electromagnetic induction, Thomson believed, could also be explained in terms of molecular vortices, as involving their inertial resistance to changes in rotational velocity. Thomson's program was sketched only very briefly and vaguely in the paper of 1856, but the paper and the program spoke to Maxwell clearly enough.[18]

"Professor Thomson has pointed out that the cause of the magnetic action on light must be a real rotation going on in the magnetic field," Maxwell wrote approvingly; the time had come to go beyond the analogical approach and begin constructing, on the basis of Thomson's picture of molecular vortices oriented along magnetic field lines, something like the "mature theory" to which Maxwell had looked forward.[19] Maxwell had begun thinking along these lines by early 1857, and he appears to have embarked on the project with considerable enthusiasm.[20] From the outset, as indicated by both the language and the substance of his letters to friends and colleagues in 1857 and early 1858, Maxwell made it clear that this was to be a theory concerning the physical nature of the field, rather than just another illustrative analogy: He was "grinding . . . at a Vortical theory of magnetism," he wrote to Cecil Monro; he was currently concerned with "the physical nature of magnetic lines of force," he wrote to Faraday; and he was designing an apparatus intended to demonstrate the reality of magnetic rotations by direct mechanical means, a drawing of which he included in a letter to Thomson.[21] Maxwell's turn, in his work on the theory of molecular vortices, to an avowed concern with – as John Herschel had put it – "the actual structure or mechanism of the universe and its parts" was thus already evident in Maxwell's communication with colleagues at the outset and was to be even more clearly exhibited in the published work that followed.

Also evident at the outset, in the correspondence with Thomson, was that Maxwell was following Thomson in seeing the idea of molecular vortices as furnishing a possible connection between electromagnetic theory and the dynamical theory of heat: Thermodynamic issues, dealing with the conversion of motion to heat and the irreversibility of this process, came up in the correspondence, and Maxwell discussed the relevance of the molecular-vortex picture to these issues in a letter to Thomson. Thomson was committed to, and Maxwell was beginning to buy into, a vision of the theory of molecular vortices as a broad, comprehensive, deep theory of the type favored at Cambridge, unifying disparate areas in the manner described in Whewell's account of "consilience"; the impetus toward unification in Maxwell's thinking was to be clear also in the published work that followed, both in terms of the unification of electromagnetic theory itself and in terms of the assimilation of optics to electromagnetic theory. There was a gestation period of about four years before any of Maxwell's thoughts concerning molecular vortices saw print; the paper "On Physical Lines of Force" then appeared in a series of four installments over a period of eleven months in 1861–2.[22]

As Maxwell indicated in the introduction to Part I of "Physical Lines," the promise of Thomson's vision, with respect to the difficulties that had hindered the development of a serious electromagnetic theory, was twofold: The theory of molecular vortices promised comprehensiveness, and it promised explanatory power. One of the major limitations of the flow analogy in "Faraday's Lines" had been its inability to give any account of the connections and interactions among electric fields, magnetic fields, and electric currents. The molecular-vortex picture, on the other hand, gave promise of just that comprehensive and connected coverage that the flow picture lacked: "If, by the [molecular-vortex] hypothesis," Maxwell

Figure 2.1. Magnetic lines of force running between unlike poles.

wrote, "we can connect the phenomena of magnetic attraction with electromagnetic phenomena and with those of induced currents, we shall have
found a theory which, if not true, can only be proved to be erroneous by
experiments which will greatly enlarge our knowledge of this part of
physics." Thus, a theory having comprehensive coverage was not guaranteed to be true, but it was, by virtue of that comprehensive coverage, to
be regarded as a serious candidate for truth; such a theory was to be
regarded in any case as having truth value, whether true or false, in
contradistinction to the flow analogy of "Faraday's Lines," which was to
be regarded as neither true nor false, but only illustrative.[23]

The molecular-vortex representation was to be distinguished from the
fluid-flow picture also in that the molecular-vortex representation was to
be regarded as having explanatory power, whereas the fluid-flow representation did not. The issue of explanatory power was a central issue for
Maxwell, and it is worth discussing at some length. Consider, for example, the case of two unlike magnetic poles, exerting mutual attractive
forces on each other (Figure 2.1). As Faraday had conceptualized this
situation in a paper entitled "On the Physical Character of the Lines of
Magnetic Force" (1852) – this is clearly the immediate referent of the title
of Maxwell's own paper – the lines of magnetic force behave as if they
have a tendency to contract along their lengths and also to repel each
other; acting in this way, they tend to pull the unlike magnetic poles
together. The attribution of this behavior to the magnetic lines of force
provides an explanation or account of the attraction between the unlike
poles, in terms of the system of lines of magnetic force existing in the
space surrounding the magnets. Faraday distinguished between a merely
geometrical treatment of the lines of force, dealing descriptively with
their distribution in space, and a *physical* treatment of the lines of force,
dealing with their dynamical tendencies that give rise to the actual forces
exerted. In the former context, "the term *magnetic line of force*" applied;
in the latter context, one spoke of a "*physical line of [magnetic] force.*"
Clearly, the titles of Maxwell's two papers reflect this usage, indicating at
the outset the merely geometrical and descriptive character of the flow

representation, as opposed to the physical and explanatory character of molecular vortices.[24]

In "Faraday's Lines," then, Maxwell had presented a mechanical representation of magnetic lines of force that adequately modeled the geometrical distribution of magnetic lines of force in space – this was given by the flow lines – but gave no purchase for understanding the forces of attraction or repulsion between magnetic poles. No tendencies of the magnetic lines to repel each other and contract along their lengths were derivable from the flow picture; consequently, magnetic forces could not be explained or accounted for by that picture. Maxwell had been explicit and insistent on this point: "By referring everything to the purely geometrical idea of the motion of an imaginary fluid, I hope to . . . avoid the dangers arising from a premature theory professing to explain the cause of the phenomena." Reviewing the matter in the introduction to "Physical lines," Maxwell stated again that in "Faraday's lines" he had been "using mechanical illustrations to assist the imagination, but not to account for the phenomena." Referring in a similar vein to Thomson's paradigmatic mechanical analogies of 1847, Maxwell observed that "the author of this method of representation does not attempt to explain the origin of the observed forces . . . but makes use of the mathematical analogies . . . to assist the imagination." Maxwell's avowed purpose in "Physical Lines," by contrast, was "to examine magnetic phenomena from a mechanical point of view, and to determine what tension in, or motions of, a medium are capable of *producing* the mechanical phenomena observed." He was seeking for a way in which "the observed resultant forces may be *accounted for.*" This was where the molecular-vortex representation showed its superiority: Assuming that the magnetic line of force represented the axis of a molecular vortex, it was easy to demonstrate that centrifugal forces would tend to make each vortex tube expand in thickness, thereby tending to increase the spacing between magnetic lines; at the same time, owing to the incompressibility of the fluid in the vortex tubes, those tubes would tend to shrink in length, giving the magnetic lines a corresponding tendency to contract along their lengths. This *physical* behavior of the magnetic lines was what was needed to *explain,* to *account for,* magnetic forces.[25]

The contrast between the illustrative mechanical analogy of "Faraday's Lines" and the explanatory mechanical theory of "Physical Lines" can be developed in a more formal manner. Consider a physical system described by a set of variables $\{F_i, G_i\}$, where the F_i are variables that represent observable mechanical forces, and the G_i are the other variables describing the system. A mechanical representation of that physical system would refer to a mechanical system represented by variables $\{f_j,\ g_j\}$, whose interrelationships were isomorphic to those of some subset $\{F_j,\ G_j\}$ of the $\{F_i,\ G_i\}$. (The larger the subset, the more complete the mechanical representation.) Whether that mechanical representation were to be construed as illustrative or explanatory would depend on the nature of the f_j: If the set $\{f_j\}$ were empty, or if the f_j were not themselves forces, then the mechanical representation would be merely illustrative; if, on the other hand, there were some f_j and they were forces, then the mechanical representation could be said to have explanatory power. In the molecular-vortex representation of "Physical Lines," for example, there are forces f_m, produced by the centrifugal forces of the rotating vortices, that are isomorphic to the forces F_m exerted in magnetostatic situations. In this case, the system of molecular vortices can be *identified with* the magnetic field, the forces f_m then being identified with the forces F_m, which is possible because they are variables of the same kind; the behavior of the molecular vortices, which explains the f_m, then also explains the F_m – that is, the theory of molecular vortices *explains* magnetic forces. In the fluid-flow representation of "Faraday's Lines," on the other hand, the set $\{f_j\}$ was empty: The variables relevant to magnetostatics, for example, were all of the g_j class – representing pole strength, field strength, and so forth – and no forces f_m corresponding to magnetostatic forces F_m were exhibited; therefore, there could be no explanation of magnetic forces by that mechanical representation – it could only be illustrative. In general, because the set $\{f_j\}$ was completely empty in the fluid-flow representation, that mechanical representation could have no explanatory power.[26]

Maxwell's assertion that the mechanical representation based on molecular vortices was explanatory rather than merely illustrative was thus no mere rhetorical device, but rather had a precise technical meaning. Because the mechanical representation based on molecular vortices was explanatory, and also because it provided comprehensive and connected coverage of the whole range of electromagnetic phenomena, Maxwell felt justified in referring to it as "the theory of molecular vortices." By calling it a theory, Maxwell indicated that it was definitely something more than a "mechanical illustration . . . to assist the imagination": It was at least a candidate for reality – perhaps "true" and perhaps "erroneous," but in any case not merely illustrative.[27]

3   *On the reality of molecular vortices*

Although the theory of molecular vortices was a candidate for reality, the strength of its candidacy was subject to vicissitudes. The four installments of "Physical Lines," taken together with other evidence from the period, furnish a rich record of the variations and nuances in Maxwell's views during the period in 1861–2 when he was working intensively on the theory. Parts I and II of the paper were published in March through May of 1861, and the nuances of Maxwell's views contained therein would appear to represent primarily a range of essentially coexisting sentiments, rather than a development in time. As discussed earlier, the tone of Part I was enthusiastic concerning the promise of the theory of molecular vortices, and guardedly optimistic concerning the theory's candidacy for reality. In Part II, however, there were crosscurrents. On the one hand, the basic "hypothesis of vortices" was characterized as a "probable" hypothesis. Also classified as probable was Maxwell's judgment concerning the sizes of the vortices: "The size of the vortices is . . . probably very small as compared with that of a complete molecule of ordinary matter." Maxwell went on to observe that although the precise sizes of the vortices could not be specified by electromagnetic measurements alone, they could be determined if one were able to measure directly the mechanical angular momentum carried by the vortices. That, in turn, "might be detected by experiments on the free rotation of a magnet," utilizing the type of apparatus that Maxwell had described in outline to Thomson back in 1858. Maxwell had indeed already "made experiments to investigate this question," but he had "not yet fully tried the apparatus" and did not report results.[28] It is clear, however, from Maxwell's continuing concern with this experiment, that in his estimation the vortices could well have been real enough to have had detectable angular momentum. He therefore continued to work on the experiment, but, as he complained to both Faraday and Thomson some months later, he had "not yet overcome the effects of terrestrial magnetism in marking the phenomenon."[29] Indeed, a successful measurement of the effect continued to elude Maxwell, as he reported retrospectively in the *Treatise on Electricity and Magnetism* (1873). His continuing efforts to detect the vortices, however, testified to their continuing candidacy for realistic status.[30]

Maxwell was not optimistic, on the other hand, concerning another part of the theory. In extending the theory to include electric currents (the procedures used to extend the theory are discussed in detail in Chapter 3), Maxwell had postulated that between the vortex cells there were interposed monolayers of small spherical particles, which rolled without slipping on the surfaces of the vortices (Figure 2.2), thus coupling the vortex rotations in the manner of "'idle wheel[s]'"; these were, moreover, to be regarded as *movable* idle wheels, and, "according to our hypothesis, an electric current is represented by the transference of the[se] moveable

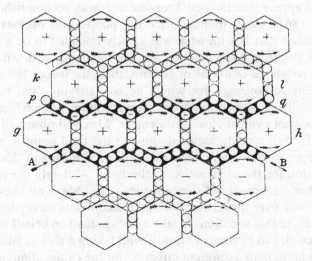

Figure 2.2. Vortex cells separated by monolayers of small spherical particles. (From Maxwell, "Physical Lines," Plate VIII, opposite p. 488.)

particles." The mathematics of these "moveable particles" turned out very neatly, but Maxwell nevertheless had to admit that the hypothesis could be regarded only as "provisional," of "temporary character." Indeed, the "conception" was "awkward," and Maxwell did "not bring it forward as a mode of connexion existing in nature, or even as that which [he] would willingly assent to as an electrical hypothesis." (Part of Maxwell's distaste for the movable particles probably stemmed from the fact that they constituted a sort of electrical fluid in his theory, and Maxwell, as a good follower of Faraday, believed in the primacy of the field and found the notion of electrical particles or fluids repugnant.) Maxwell, then, was careful to distinguish between a part of the theory that he regarded as a "probable" hypothesis and a good candidate for reality, and another part, which he regarded as a "provisional" hypothesis – a kind of placeholder in the theory – and a very poor candidate for reality. (Herschel, for example, had allowed for both kinds of hypotheses in his theory of scientific method.[31])

Part II of "Physical Lines" ended on a down note. As I shall argue in Chapter 3, Maxwell encountered difficulty in extending his theory to embrace electrostatics, and he did not anticipate, when he submitted Part II of the paper for publication, the triumphant further extension of the theory that he was to publish as Parts III and IV after an eight-month hiatus. (It would appear that the new breakthrough was accomplished in the summer of 1861, some months after the publication of Part II.) Maxwell thus closed Part II with what was clearly intended to be a final conclusion to the paper, and he evidently was not too happy with what he had wrought. Through the omission of electrostatics, the theory fell short of the comprehensive coverage for which he was striving, and he remained still at a disadvantage vis-à-vis Weber's more comprehensive theory. In a somewhat cynical closing statement, Maxwell observed that "those who [had] been already inclined" toward the field point of view might find his paper worthwhile. Others might be predisposed to "look in a different direction for the explanation of the facts" – clearly the reference was to Weber's action-at-a-distance theory – and Maxwell knew he was not going to win over those others with the incomplete theory he had presented. Unable, in that situation, to take a strong stand on behalf of his own theory, Maxwell had to content himself with taking a shot at Weber's theory: "Those who look in a different direction for the explanation of the facts, may be able to compare this theory with that . . . which supposes electricity to act at a distance with a force depending on its velocity, and therefore not subject to the law of conservation of energy." Thus, at that point, Maxwell was using the dynamical theory of heat and the associated conservation law not only as one of the foundations of his own theory but also as a normative principle on the basis of which to criticize alternative theories.

Continuing in a critical and dyspeptic vein, Maxwell summed up his own contribution as follows: "We have now shewn in what way electromagnetic phenomena may be imitated by an imaginary system of molecular vortices." This statement is clearly out of tune with Maxwell's stance in the rest of "Physical Lines"; it echoes, instead, Maxwell's tone of diffidence with respect to Weber in "Faraday's Lines." Maxwell had hoped to construct a theory that would rival Weber's in comprehensiveness; he had failed, and he was disappointed. Apparently, however, his disappointment spurred him on to further efforts, which were crowned with spectacular success.[32]

Part III of "Physical Lines" was published in January 1862, and there, by extending the theory to electrostatics and by providing an explanation of electrical forces in terms of stresses in the medium, Maxwell finally achieved the full comprehensiveness and explanatory character that he had sought. He was now able to "explain the condition of a body with respect to the surrounding medium when it is said to be 'charged' with electricity, and account for the forces acting between electrified bodies, [thereby] establish[ing] a connexion between all the principal phenomena of electrical science." The extension to electrostatics was accomplished by assigning elastic properties to the molecular vortices, making the system of vortices in space – the "magneto-electric medium" – capable of sustaining elastic waves. Using values of electrical parameters measured by Wilhelm Weber and Rudolph Kohlrausch – Weber was at this point being enlisted as an ally – Maxwell was able to calculate the velocity of elastic waves in the magnetoelectric medium, arriving at a value that agreed precisely with (in fact was bracketed by) existing measurements of the speed of light in air or vacuum. Maxwell's conclusion – the route to which will be discussed in detail in Chapter 5 – was that "we can scarcely avoid the inference that *light consists in the transverse undulations of the same medium which is the cause of electric and magnetic phenomena.*" Maxwell's use of italics was well justified: he, and soon his colleagues, judged this to be a result of immense consequence, and historical perspective has reinforced that assessment (see further Chapter 6).[33]

The result was also of the greatest importance as concerned the status of the theory of molecular vortices. First, given the widespread belief – at Cambridge and beyond – in the real existence of the luminiferous medium, the identification of the luminiferous and magnetoelectric media became an argument for the real existence of the magnetoelectric medium, with its vortex structure. At the level of particulars, certain features of the vortices – their elasticity, for example – were no longer ad hoc to electromagnetic theory, but rather could be seen as the natural extension of a theoretical structure already in place, namely, the wave theory of light. Relatedly, the generality and broad range of the resulting unified theory of electromagnetism and optics argued powerfully for the physical significance of the theory of molecular vortices. Indeed, this was a most spectacular example of Whewellian consilience. Moreover, as Maxwell stressed in a letter announcing the new results to Faraday, the new, unified theory predicted various relationships between electromagnetic and optical phenomena, experimental verification of which would help to establish the truth value of the theory by demonstrating its predictive power (a criterion emphasized by both Herschel and Whewell): "The conception I have hit on has led, when worked out mathematically, to some very interesting results, capable of testing my theory, and exhibiting numerical relations between optical, electric, and electromagnetic phenomena, which I hope soon to verify more completely." In a parallel letter to Thomson, Maxwell acknowledged Thomson's seminal role in the development of the theory of molecular vortices, sketched the application of the theory of molecular vortices to electricity and magnetism, making clear the comprehensiveness and explanatory character of the theory, and discussed the experimental basis and implications of the unification with optics. In all three announcements of the new results – the public one in "Physical Lines" and the private ones to Faraday and Thomson – Maxwell's confidence in the theory of molecular vortices, now with a vastly enhanced range of applicability, was in marked contrast to his earlier vacillations. Further arguments for the reality of molecular vortices were to be presented in the final installment of "Physical Lines," which appeared one month later.[34]

4 The mathematics and physics of linear and rotatory vectors

Maxwell's most focused argument in favor of the reality of molecular vortices was given in Part IV of "Physical Lines." It was a mathematical argument, but it was not presented as pure and abstract mathematics; it was, rather, according to Maxwell's enduring commitment, "embodied mathematics" – mathematics represented in mechanical examples.[35] The central mathematical relationship on which Maxwell focused was that exemplified in the relationship between electric current and magnetic field, that is, Ampère's circuital law in differential form:

$$p = \frac{1}{4\pi}\left(\frac{d\gamma}{dy} - \frac{d\beta}{dz}\right)$$

$$q = \frac{1}{4\pi}\left(\frac{d\alpha}{dz} - \frac{d\gamma}{dx}\right) \qquad (2.1)$$

$$r = \frac{1}{4\pi}\left(\frac{d\beta}{dx} - \frac{d\alpha}{dy}\right)$$

where α, β, γ are the Cartesian components of the vector representing magnetic field intensity, p, q, r are the components of the vector representing electric current density, and the differential operators d/dx, d/dy, d/dz represent partial differentiation with respect to Cartesian coordinates x, y, z (Maxwell did not use a special symbol for partial differentiation). In order to illustrate the deeper meaning of this equation, Maxwell listed a series of mechanical examples to which this equation could be applied: (1) If α, β, γ represents linear displacement or change of location, then p, q, r represents rotatory displacement or change of location. (2) If α, β, γ represents linear velocity, then p, q, r represents rotational velocity. (3) If α, β, γ represents a force or push, then p, q, r represents a torque or twist. Common to these examples is the circumstance that α, β, γ has "*linear* . . . character," representing a motion or a thrust in a certain direction, whereas p, q, r has "*rotatory* character," representing a rotation or a twist about a certain axis. Further mechanical examples exhibited the inverse relationship: (4) If α, β, γ represents rotatory displacement or motion in a continuous medium, then p, q, r represents linear displacement or relative motion in that medium. (5) If α, β, γ represents the rotational velocities (angular velocities) of the vortices in Maxwell's theory, then p, q, r represents the averaged linear flow density of the idle-wheel particles postulated in that theory. In these examples, it is α, β, γ that represents a rotatory motion about some axis, whereas p, q, r represents an associated linear motion. Maxwell's conclusion from the two sets of instances was that equations (2.1) in general represent a kind of relationship that obtains "between certain pairs of phenomena, of which one has a *linear* and the other a *rotatory* character"; if α, β, γ is linear, then p, q, r is rotatory, and if α, β, γ is rotatory, then p, q, r is linear.[36]

The content and character of Maxwell's argument can be highlighted by comparison and contrast with a modern treatment of the same issue. Modernly, and in fact building on Maxwell's continuing work in this area, one maintains his distinction between two kinds of vectors: Maxwell's linear quantities are now denoted true vectors or polar vectors, and Maxwell's rotatory quantities are now designated pseudovectors or axial vectors. The distinction between these two kinds of vector quantities is, however, established in a modern treatment on the basis of their transformation characteristics – in particular with respect to spatial reflection – rather than by reference to mechanical examples. Maxwell, conversely, relied exclusively on mechanical examples to establish the distinction and to illustrate the relationship between the two kinds of vectors, as in equations (2.1).[37] Maxwell was, after all, a mechanical philosopher: not a mechanical philosopher of the eighteenth-century type, with their qualitatively distinct subtle matters, but a characteristic British mechanical philosopher of the post-1850 period, who espoused an ontology of matter and motion in which "all matter must in itself be the same, and can be modified only by differences of arrangement and motion and by being actuated by different systems of force." Given that ontology, the reality underlying the electromagnetic field had to be mechanical, and the equations describing electromagnetic relationships, ultimately, were the expression of mechanical conditions. Mechanically, equations (2.1), by virtue of their mathematical structure, had to represent the relationship between a linear mechanical motion or force and a rotational motion or torque. Either magnetism was rotational and electric current linear, or magnetism was linear and electric current rotational. The mathematical relationship of equations (2.1), taken together with the mechanical ontology, guaranteed this; all that remained was to make a decision between the two possibilities.[38]

The way to tell whether a given phenomenon had linear or rotatory character, Maxwell suggested, was to look at its effects: "All the direct effects of any cause which is itself of a longitudinal character, must themselves be longitudinal, and . . . the direct effects of a rotatory cause must be themselves rotatory." Maxwell proceeded to inventory the effects of electric currents, to judge whether they were to be classified as linear or rotatory in character. In the first place, Maxwell observed, "electric currents are known to produce effects of transference in the direction of the current." In the electrolysis of water, for example, hydrogen is moved in one direction along the current line, and oxygen in the other, thus indicating a linear character for the electric current; there was no known rotational effect of electric current (Faraday had in fact searched for such an effect but found none), so that Maxwell felt confident in characterizing the electric current as linear. Given the previous argument concerning equations (2.1), this result was sufficient to demonstrate not only that electric current was linear in character but also that magnetism was rotatory; any additional information bearing directly on the rotatory character of magnetism would then introduce a reassuring redundancy into the argument.[39]

As directly concerned the nature of magnetism itself, Maxwell first argued that magnetism produced no known linear effects. (The magnetic lines of force and their actions on magnetic poles appear to be linear, but, as Maxwell argued, some phenomenon, such as electrolysis in the electrical case, where the opposite ends of the line of action are physically distinguished, is needed to establish that the line is not merely an axis of rotation; no such phenomenon, however, had been found for the magnetic line of force. This is a subtle point, which still bedevils students.) Magnetism did, however, produce a rotatory effect, namely, the Faraday rotation – "the rotation of the plane of polarized light when transmitted along the lines of magnetic force." That, of course, was the phenomenon that had provided the basis for the whole line of thinking that Maxwell was pursuing. Maxwell acknowledged Thomson's role in "point[ing] out that the cause of the magnetic action on light must be a real rotation going on in the magnetic field." Also invoked was Thomson's argument concerning the handedness of the Faraday rotation, leading to the conclusion that "the direction of rotation is directly connected with that of the magnetic lines, in a way which seems to indicate that magnetism is really a phenomenon of rotation."[40]

What Maxwell had added to Thomson's original argument for the reality of magnetic rotations was a fourfold redundancy: Given Maxwell's argument concerning the mathematical character of equations (2.1) and their associated mechanical significance, four kinds of evidence converged on the conclusion that magnetism was rotational: (1) linear effects of electric current, as in electrolysis; (2) lack of rotatory effects of electric current, as generally established and as further tested in Faraday's experiments; (3) lack of linear effects of magnetism, as generally established and as supported by Maxwell's argument concerning the lack of any physical distinction between the two ends of a magnetic line of force; and (4) rotatory effects of magnetism, as established experimentally by Faraday and as interpreted theoretically by Thomson. Maxwell added further redundancy to the argument in favor of the rotational character of magnetism by enlisting even the views of the action-at-a-distance theorists – Ampère and Weber, in particular – to the effect that electric currents involved linear transport of electric charge, whereas magnetism was a manifestation of current loops.[41]

The strength and weakness of this argument in Part IV of "Physical Lines" for the rotatory character of magnetism was its generality, its independence of the specifics of the theory of molecular vortices. The details of the theory of molecular vortices were not supported by this argument, and where those details were questionable – as in the case of the idle-wheel particles – they remained questionable. On the other hand, as this argument was not tied to those details, it could retain its general force for Maxwell even when he backed away from the details. In fact, as will become apparent in the sequel, Maxwell never did give up the belief that there was "a real rotation going on in the magnetic field." Given the mechanical ontology, the mathematics of linear and rotatory vector quantities, and the experimental evidence concerning the linear and rotatory characteristics, respectively, of electric currents and magnetic fields, the conclusion for "a real rotation" was simply unavoidable for Maxwell.[42]

5   *The decline of the theory of molecular vortices*

After the publication of "Physical Lines," Maxwell began a measured retreat from the mechanical concreteness and detail that characterized his presentation of the theory of molecular vortices there. His next major paper on electromagnetic theory, "A Dynamical Theory of the

Electromagnetic Field" (1864–5), exemplified this trend in his thinking. Clearly, dissatisfaction with the weakest link in the theory – the idle-wheel particles – was a crucial reason for this retreat. There were, however, other reasons, having to do with the broader development of Maxwell's research program: His research on gas theory, in the 1860s, took a direction that had negative implications concerning molecular vortices, and his concern to develop the electromagnetic theory of light in a manner that would be acceptable to a broader audience led also to a deemphasis on molecular vortices. The result of these was to redirect Maxwell's efforts in electromagnetic theory toward a more phenomenological emphasis, without, however, engendering either a return to the analogical approach or a complete loss of faith in the reality of molecular vortices.[43]

The original stronghold of the hypothesis of molecular vortices had been the theory of heat and gases, and one of the most attractive features of Thomson's suggestion, in 1856, that the hypothesis of molecular vortices be applied to electromagnetic phenomena was the broad unification this promised. In the later 1850s, however, most notably as a result of the work of Rudolph Clausius, the idea that the motion of gas molecules giving rise to gas pressure was translatory – characteristically in straight lines – rather than rotatory began to look more and more appealing. Maxwell's initial response to this situation, in the years around 1860, was to maintain his primary allegiance to molecular vortices, employing them as the basis for a physical theory in "Physical Lines," while utilizing the linear-motion picture as the basis of a physical analogy in gas theory, with no commitment to it as a realistic representation of nature. In this way, conflict between the respective requirements of electromagnetic theory and gas theory was minimized. This accommodation, however, proved to be unstable.[44]

In the course of the 1860s, Maxwell's investment in and commitment to the linear picture – that is, to what has been ever since the standard kinetic theory of gases – increased substantially. Exhibiting the same kind of progression from an analogical stage to a theoretical stage that we have already seen in his electromagnetic theory, Maxwell in 1866 published what he was prepared to call a "Theory of Gases"; once again, just as in the electromagnetic case, the step to the theoretical stage was justified by the comprehensiveness and explanatory character of the given mechanical representation. The methodological progression in Maxwell's gas theory thus mirrored quite faithfully the methodological development of his electromagnetic theory; as far as content was concerned, however, the direction in which Maxwell's gas theory was developing in the first half of the 1860s tended to undermine the foundations of his electromagnetic theory. Not that there was any direct conflict between the gas theory and the electromagnetic theory – molecular vortices in the ether were perfectly consistent with translational motion of gas molecules – but the grand synthesis on the basis of molecular vortices as originally envisaged by Thomson appeared to have been ruled out, and some of the appeal of applying molecular vortices to electromagnetic phenomena was thereby lost.[45]

In part, it was the very success of the theory of molecular vortices that led to its downfall. This theory had provided the context for a unified treatment of electromagnetism and optics on the basis of the mechanics of one universal medium or ether. The further development and consolidation of this unification of electromagnetism and optics then became the new focus of Maxwell's continuing research. On the experimental side, he became quite productively involved in work on electrical measurements and standards, oriented toward more precise determination of the ratio of electrical units, on the basis of which the connection with light had been established. On the theoretical side, he worked toward the establishment of a more direct connection between electromagnetism and optics. The original connection between the two, as established in "Physical Lines," has aptly been designated an "electro-mechanical"[46] theory of light, rather than an electromagnetic theory of light: In "Physical Lines," Maxwell had argued from electromagnetic phenomena, by way of the theory of molecular vortices, to the mechanical properties of the magnetoelectric medium; then, from the mechanical properties of that medium, Maxwell had deduced the propagation of transverse elastic waves in it, with all the characteristic properties of light waves. Early on, however, Maxwell surmised that a more direct and theoretically parsimonious establishment of this result should be possible, and a more direct argument, "cleared . . . from all unwarrantable assumption," was likely to be more palatable to the Continental action-at-a-distance electricians, who were skeptical of the whole field-primacy approach of Faraday, Thomson, and Maxwell. Maxwell was in fact successful in devising a more parsimonious argument, which proceeded directly from the electromagnetic equations – as appropriately modified – to the calculation of electromagnetic waves propagating at the velocity of light. This, finally, was truly an "electromagnetic theory of light," and it formed the centerpiece of Maxwell's paper "A Dynamical Theory of the Electromagnetic Field."[47]

In this paper, then, in consonance with the exigencies of his research programs in both gas theory and electromagnetic theory, Maxwell retreated from the specifics of the theory of molecular vortices, but not from the general framework. He still insisted on the existence of a mechanical medium in space that was both the carrier of light waves and the seat of electric and magnetic fields; he was still willing to propose, as a "very probable hypothesis," that magnetic and electric fields were manifestations respectively of "motion" and "strain" in that medium; and he still judged that the Faraday rotation gave "reason to suppose that th[e] motion [underlying the magnetic field was] one of rotation, having the direction of the magnetic force as its axis." Beyond this Maxwell was not willing to go, and he did not even use all of this in developing the mathematical theory. Basically, all that Maxwell used were the equations of electromagnetic phenomena as established by experiment, together with the assumption that these reflected conditions in a connected mechanical medium pervading space and capable of storing, exchanging, and transmitting kinetic and potential energy. This warranted Maxwell in treating the field variables (and other electromagnetic variables) as generalized mechanical variables, in the sense of the Lagrangian formalism. The result was a "dynamical theory" of the electromagnetic field, which was still a mechanical theory, but abstract and general rather than concrete and pictorial.[48]

6   *Molecular vortices in the* Treatise on Electricity
        and Magnetism

Maxwell's primary methodological commitment in the later 1860s and the 1870s was to the dynamical approach, which abjured completely the use of concrete mechanical images; the dynamical approach was fully developed and centrally positioned in the *Treatise on Electricity and Magnetism* (1873). The *Treatise*, however, was intended to be a comprehensive work, treating all aspects of electromagnetic phenomena, and some of these were not amenable to a treatment by macroscopic dynamical theory, without any assumption as to mechanical details. In particular, an entire chapter was devoted to the Faraday rotation (important because of its bearing on the electromagnetic theory of light), and here molecular vortices played a central role. Maxwell began his analysis – characteristically for the *Treatise* – by "consider[ing] the dynamical condition[s]" attendant upon the Faraday rotation, that is, by applying the Lagrangian formalism; he arrived thereby at the following – by now familiar – result, stated in the abstract terminology characteristic of that formalism: "the consideration of the action of magnetism on polarized light leads . . . to the conclusion that in a medium under the action of magnetic force something belonging to the same mathematical class as an angular velocity, whose axis is in the direction of the magnetic force, forms a part of the phenomenon." He then proceeded to interpret this result in the light of his mechanical ontology: Given a universe of matter and motion, the "something belonging to the same mathematical class as an angular velocity" must in fact *be* an angular velocity of some rotating portion or portions of the medium filling space. Experiment indicated that no sizable angular momenta were associated with these rotations, so the rotating portions of the medium had to be small, and the conclusion was that "we must therefore conceive the rotation to be that of very small portions of the medium, each rotating on its own axis." "This," once again, was "the hypothesis of molecular vortices"; Maxwell used the hypothesis basically in this unadorned form, not fleshing it out as he had in "Physical Lines."[49]

In a final summary of his ultimate views concerning the reality of molecular vortices, Maxwell again invoked the distinction that he had made in "Physical Lines" between the status of the vortices themselves – they had been a "probable" hypothesis – and the status of the system of idle-wheel particles that coupled the motions of the vortices – this had been a "provisional," "temporary," and "awkward" hypothesis. Concerning the vortices themselves, the "probable" hypothesis, Maxwell had this to say in the *Treatise:* "I think we have good evidence for the opinion that some phenomenon of rotation is going on in the magnetic field, that this rotation is performed by a great number of very small portions of matter, each rotating on its own axis, this axis being parallel to the direction of the magnetic force, . . ." Concerning this part of the theory of molecular vortices, then, Maxwell was fully as sanguine in the *Treatise* as he had been in "Physical Lines"; his language in the *Treatise* – "I think we have good evidence for the opinion that . . ." – was, if anything, a bit stronger than in "Physical Lines."[50]

Concerning the *existence* of a mechanism coupling the motions of the individual vortices, there was also good evidence: "I think we have good evidence . . . that the rotations of the . . . different vortices are made to depend on one another by means of some kind of mechanism connecting them." The *particular* connecting mechanism envisaged in "Physical Lines" – that is, the system of idle-wheel particles – was, on the contrary, not to be taken seriously:

> The attempt which I then made to imagine a working model of this mechanism must be taken for no more than it really is, a demonstration that mechanism may be imagined capable of

> producing a connexion mechanically equivalent to the actual
> connexion of the parts of the electromagnetic field. The prob-
> lem of determining the mechanism required to establish a
> given species of connexion between the motions of the parts of
> a system always admits of an infinite number of solutions.

This agnostic statement notwithstanding, Maxwell observed that certain things were more definitely known even about the connecting mechanism: "Electromotive force arises from the stress on the connecting mechanism [and] electric displacement arises from the elastic yielding of the connecting mechanism." Thus, certain *general* mechanical properties of the connecting mechanism were known, along with their relationships to electromagnetic phenomena.[51]

An infinite number of different mechanisms could be imagined that would fulfill these specifications, and Maxwell clearly did not entertain the hope that one could ever determine which of these was the "actual connexion" existing in nature. The effort that Maxwell had made in "Physical Lines" to envisage a concrete example of such a mechanism had not, however, been a worthless exercise; it had provided a "demonstration that mechanism may be imagined capable of" fulfilling the given specifications. Such a concrete mechanism, not realistically intended, but intended instead to show that a mechanism of the sort required was possible, was called by Maxwell a "working model." A working model is similar to a physical analogy in that it is a concrete and pictorial mechanical representation, with imaginary rather than realistic status. There is, however, an important difference: The working model must be able to *produce* the mechanical effect in question – that is, it must be a model that really *works*, in the sense of accomplishing the effect. The working model furnishes a possible explanation of the effect, just because it is able to produce the effect. One may judge it highly improbable that the given working model faithfully represents the details of the actual situation, either because the working model is manifestly awkward or artificial, or simply because one knows that there is an infinite number of possible working models, so that the a priori probability that a given one is the true one is vanishingly small. Nevertheless, the working model is "capable of producing" the observed effects, and in that sense represents a *possible* explanation. A physical analogy, however, will in general not represent even a possible explanation, because the mechanical system envisaged is not capable of producing the phenomenon in question. (In the formal language introduced earlier, a working model is described by variables $\{f_j\}$ that are forces, whereas a physical analogy is not.) Thus, even as concerned the connecting mechanism – the weakest part of the theory of molecular vortices – Maxwell was not retreating back to the physical analogy stage.[52]

To sum up Maxwell's final position concerning the theory of molecular vortices, he regarded part of the theory as a hypothesis for which "we have good evidence," and part of the theory as a "working model." In order to get a comprehensive theory of electromagnetic phenomena, one would have to put the two parts together; evaluating the resultant theory by its weakest link, one would have to characterize the whole as merely a "working model." Maxwell, however, chose to maintain the separation of the two parts in his characterization of the status of the theory, thus highlighting the strong and continuing commitment he had to the reality of the central core of the theory, the vortices themselves.[53] In addition, Maxwell judged certain general mechanical features of the theory to be firmly established throughout, including both the vortices and the connecting mechanism.

The following results of the theory, however, are of higher value:

(1) Magnetic force is the effect of the centrifugal force of the vortices.

(2) Electromagnetic induction of currents is the effect of the forces called into play when the velocity of the vortices is changing.

(3) Electromotive force arises from the stress on the connecting mechanism.

(4) Electric displacement arises from the elastic yielding of the connecting mechanism.[54]

Maxwell's popular lectures and writings through the 1870s continued to echo his judgment in the *Treatise* that the core of the molecular-vortex theory was still sound. In a talk at the Royal Institution "On Action at a Distance," Maxwell argued that "strict dynamical reasoning" demonstrated the existence of "molecular vortices . . . rotating, each on its own axis," with "magnetic force[s] . . . aris[ing] from the centrifugal force of the . . . vortices." In his article "Ether" for *Encyclopaedia Britannica*, Maxwell's message was similar: "Sir W. Thomson has shewn" that there is "a rotational motion in the medium when magnetized," which "must be a rotation of very small portions of the medium each about its own axis, so that the medium must be broken up into a number of molecular vortices." Finally, in the article "Faraday" for *Britannica*, Maxwell observed that the discovery of the Faraday rotation, though it had not led to much in the way of "practical application," had nevertheless been "of the highest value to science, as furnishing complete dynamical evidence that wherever magnetic force exists there is matter, small portions of which are rotating about axes parallel to the direction of that force." To the end, then, Maxwell maintained his allegiance to the image of whirling vortices as the basis of the magnetic field.[55]

Conclusion

Viewed over the course of his entire career as a theorist of electricity and magnetism, Maxwell's use of mechanical representation was varied and pluralistic, reflecting the various formative influences on his work as well as the developing needs of his research program. The Scottish emphasis on analogy, rooted in a combination of Baconian empiricism and Common Sense philosophical sophistication, and transmitted through James D. Forbes, William Hamilton, and William Thomson, was evident in Maxwell's initial use of the method of physical analogy. Then, with growing confidence in the foundation of his work, Maxwell embraced the hypothetical method of John Herschel, William Whewell, and the Cambridge school; in this context, Maxwell utilized the hypothesis of molecular vortices as the foundation for a broad and ramified "physical theory" of electricity and magnetism, put forward with considerable realistic intent. Finally, there set in a period of disillusionment concerning at least certain aspects of the theory of molecular vortices, resulting in a measured retreat from that kind of detailed and explicit mechanical theorizing; in this final phase of Maxwell's work, a limited and nuanced continuing reliance on certain aspects of the theory of molecular vortices coexisted with a predominant emphasis on a more abstract variety of mechanical theory – based on the Lagrangian formalism – that completely renounced the use of concrete mechanical images. If one were to look only at Maxwell's starting and ending points, one would conclude that he took primarily a skeptical stance toward mechanical representation. That, however, would be to ignore the middle period, which was the period of strong mechanical commitment, and also the period of intensive innovation in Maxwell's electromagnetic theory.

Maxwell's turn away from concrete mechanism in the later 1860s and 1870s – echoing his initial skepticism of the 1850s – was of great consequence for the subsequent methodological and foundational development of physical science: Maxwell's turn away from mechanical models was one of the precipitating events in the decline of the mechanical worldview and the transition to the more abstract physical formalisms of the twentieth century. Consequently, much historical and philosophical analysis has emphasized this skeptical element in Maxwell's approach to mechanical models.[56] The other aspect of Maxwell's approach to mechanism – the strong commitment to the molecular-vortex model in the middle period – has received less attention. Indeed, viewed from a twentieth-century perspective, the molecular-vortex model may appear to be "bizarre" and "outlandish," and hence not worthy of serious attention.[57] Maxwell, however, took that model very seriously at the time, and so must we if we are to fully understand his work. In particular, the two major innovations in Maxwell's electromagnetic theory – the displacement current and the electromagnetic theory of light – received their initial formulations in the context of the molecular-vortex model. If we are to understand the origins of these crucial novelties, in terms of the context of nineteenth-century mechanical commitment that gave birth to them in Maxwell's work, it will be necessary to look more closely at the details of the construction of the molecular-vortex model.

3

The elaboration of the molecular-vortex model

As we have seen, Maxwell took the molecular-vortex model quite se-riously – with ontological intent – when he first presented it in 1861–2, and although he later lost confidence in certain aspects of the model and removed it to the periphery of his research program, he continued in his allegiance to the core hypothesis of the model – that is, to the hypothesis of molecular vortices. The centrality of the molecular-vortex model in Maxwell's general thinking about electromagnetic theory and the particu-lar importance of this model in the background of the displacement cur-rent and the electromagnetic theory of light together provide motivation for a careful study of this intricate mechanical model of the electromagne-tic field.

Maxwell's work on the molecular-vortex model was guided, above all, by his desire – his commitment – to fashion a coherent and comprehen-sive theory unifying the full range of electromagnetic phenomena from the field-primacy point of view. This was required in order to produce a credible alternative to Wilhelm Weber's unification of electromagnetic theory within the charge-interaction framework; comprehensiveness and coherence were required also in connection with the intended realistic status of the theory – Scottish and Cambridge methodologies converged on this requirement. Constructing such a theory was, however, not a straightforward task, and the molecular-vortex model did not spring full-blown from Maxwell's imagination at some given moment in time; in-stead, the model developed over a period of years, beginning in 1857, and was published in a series of installments over an eleven-month period in 1861–2. To understand the molecular-vortex model is to understand the process of its construction – that is, the process of its development and elaboration over time, in such a manner as to ensure coherence and consistency throughout.[1]

There are two senses in which one can talk about the *development* of the theory: First, there is the rhetorical development of the theory, as it unfolded in the four installments of "Physical Lines." Second, there is the historical development of the theory, as it emerged in the successive stages of Maxwell's thinking on the subject. In some cases, the historical development and the rhetorical development of a theory may be quite unrelated, as, for example, when the rhetorical development is axiomatic, whereas the historical development was inductive, or when the rhetorical development is inductive, whereas the historical development was hypothetical. In the case of the theory of molecular vortices, however, there is strong evidence that the historical development of the theory and its rhetorical development, as presented in "Physical Lines," are closely related. The rhetorical development of the theory can be apprehended on the basis of its explicit presentation in the text of "Physical Lines"; the historical development of the theory, on the other hand, is something that cannot be read from a text, but must be inferred, using various evidence, including but not limited to the text of "Physical Lines." The strategy of this chapter is to work through the successive installments of "Physical Lines," interpreting on both levels, rhetorical and historical, and making use of both internal and external evidence to distinguish the two.

Section 1 deals with the original core of the theory – the basic molecular-vortex model as applied to magnetostatics. Sections 2–4 deal successively with the series of elaborations of the model that Maxwell undertook in order to embrace in the theory the phenomena of electromagnetism, electromagnetic induction, and electrostatics – thereby ultimately achieving the kind of comprehensive coverage that had been his aim throughout. In order to ensure that the coherence of the theory would be maintained as it was extended and elaborated, Maxwell required that each newly introduced variable satisfy a criterion of proper linkage with the established core of the theory; his employment of this criterion in the successive stages of the extension of the theory is analyzed in schematic form in displays (3.7), (3.10), (3.11), and (3.15), representing, respectively, the extensions of the theory to embrace electric current, electromotive force, electric charge, and the electrostatic field.

1   *Molecular vortices applied to magnetic phenomena:*
       the core of the theory

       Part I of "Physical Lines," subtitled "The Theory of Molecular Vortices Applied to Magnetic Phenomena," was published in March of 1861. In making the application of the theory to magnetism the rhetorical starting point of his presentation, Maxwell was echoing the historical reality. As Maxwell clearly acknowledged – many times over – the idea of applying the concept of molecular vortices to electricity and magnetism had come from William Thomson, and Thomson's primary application of the molecular-vortex picture had been in the areas of magnetism and magnetooptics. Thomson had looked forward to an explanation of magnetic forces in terms of the pressures that would exist in a system of molecular vortices geometrically arranged so as to correspond to the magnetic field, and Maxwell devoted Part I of "Physical Lines" to an investigation of such pressures, attributable to the centrifugal forces of the vortices, which he then interpreted as giving rise to magnetic forces.[2]

In certain other respects, however, Maxwell's rhetoric at the beginning of "Physical Lines" was not at all faithful to the historical reality: He did not make any allusion whatever to the origin of the hypothesis of molecular vortices in Rankine's version of the kinetic theory of heat, he did not mention the Faraday rotation, and he did not discuss Thomson's argument for connecting Rankine's vortices with the Faraday rotation. Indeed, Rankine's kinetic theory was not mentioned anywhere in the paper (although Rankine was cited for the formalism of the representation of mechanical stresses); the Faraday rotation was brought up only in Part IV of the paper; and Thomson's dynamical argument establishing a connection between vortical motion in the medium and the Faraday rotation was also brought up only in Part IV of the paper – whose publication had not even been anticipated when Maxwell published Part I. (Thomson's elastic-solid analogy of 1847 as applied to magnetism was mentioned in Part I of "Physical Lines," but only by way of contrast between Thomson's geometrical analogy and Maxwell's explanatory theory.) Maxwell's failure to call attention to the connection with Rankine's kinetic theory is understandable: By 1861, Rankine's theory of molecular vortices, as an account of the nature of heat, no longer looked as promising as it once had. As discussed in Chapter 2, the work of Rudolph Clausius and others in the later 1850s had provided strong support for the idea that heat was a manifestation of linear rather than rotatory motion at the microlevel, and Maxwell himself had exploited the linear-motion picture in a paper of 1859–60 on gas theory. Maxwell, therefore, in 1861, apparently preferred not to associate Rankine's theory of heat with his own theory of magnetism.[3]

Maxwell's omission of any mention of the Faraday rotation and Thomson's treatment of it is, at first look, more puzzling. In Part IV of the paper, Maxwell was to be clear enough about his debt to Thomson's analysis of the Faraday rotation, and that debt was also clearly acknowledged, even before publication of Parts III and IV of the paper, in a letter to Thomson in which Maxwell characterized the origin of the theory of molecular vortices as follows:

> Since I saw you I have been trying to develope the dynamical theory of magnetism as an affection of the whole magnetic field according to the views stated by you in . . . 1856 . . . 1857 . . . and elsewhere.

Given this admitted indebtedness to Thomson, the lack of acknowledgment of his work in Part I of "Physical Lines" may have come across as ungracious, and one wonders if that may have been an element in Thomson's subsequent coolness toward some aspects of Maxwell's theory. (On the other hand, scientific citation practices in the nineteenth century in general, and Maxwell's practice in "Physical Lines" in particular, were oriented more toward supporting the argument in progress than toward acknowledging intellectual debts and giving credit to forerunners; that Maxwell did not feel constrained to cite either Rankine or Thomson can be better understood against this background.[4])

Considered as a rhetorical move per se, Maxwell's decision not to mention the Faraday rotation and Thomson's treatment of it at this point can be seen to have had certain merits. Basically, Maxwell's introductory gambit was phrased in terms of the field approach versus the action-at-a-distance approach:

> We are dissatisfied with the explanation [of magnetic force] founded on the hypothesis of attractive and repellent forces directed towards the magnetic poles, even though we may have satisfied ourselves that the phenomenon is in strict accordance with that hypothesis, and we cannot help thinking that in every place where we find these lines of force, some physical state or action must exist in sufficient energy to produce the actual phenomena.[5]

In this context, the Faraday rotation did not constitute a crucial experiment, because it could be explained in either the charge-interaction or the field-primacy framework: Circular or rotatory molecular currents were postulated in the theories of Ampère and Weber, and they could be applied to an understanding of the Faraday rotation; molecular vortices supplied an alternative explanation from the field point of view. Thus, the Faraday rotation was not relevant to Maxwell's primary rhetorical purpose in Part I of "Physical Lines," namely, to promote the field approach to electricity and magnetism. Furthermore, the magnetoelectric medium was pictured by Maxwell in Part I of "Physical Lines" as a fluid medium,

not evidently capable of propagating light waves, and not evidently connected in any way with the luminiferous ether; any discussion of the
connection of magnetism with optics, therefore, would have been without
proper basis at that point in the presentation and therefore was best left
out.[6]

In the absence of the inductive support for magnetic rotations that
could have been supplied by the Faraday rotation, Maxwell was left with
a classical hypothetico-deductive rhetorical structure for Part I of "Physical Lines": The hypothesis of molecular vortices was stated with little
inductive motivation; then, as Maxwell himself described it, "the mechanical consequences of [the] hypothes[i]s" were drawn, and "the
. . . results" were "compar[ed] with the observed phenomena of magnetism." It was a type of argument that was quite familiar in the Cambridge milieu (cf. the discussion of Herschel in Chapter 1, Section 3), and
Maxwell developed the deductive part as a set of mathematical propositions and demonstrations, very much in the spirit of the Cambridge mixed
mathematics.[7]

In general outline, the derivation of magnetic forces from the hypothesis of molecular vortices proceeded as follows: The "magneto-electric
medium" was to be conceived as a fluid medium; in a region of nonzero
magnetic field, this medium would be filled with innumerable small
vortex tubes or filaments, corresponding in geometrical arrangement to
the magnetic field lines; the angular velocities of these "molecular vortices" would be taken proportional to the field intensity. The rotational
motions thus envisioned would engender centrifugal forces, causing the
vortex filaments to have a tendency to expand equatorially and contract
along their lengths; the corresponding magnetic field lines would appear
to repel each other and have a tendency to shorten. This was exactly the
behavior of the magnetic field lines that had been discussed earlier by
Faraday and was just what was needed to account for magnetic forces: In
the case of opposite magnetic poles (returning to the example of Figure
2.1), where the lines of force run from one pole to the other, this kind of
stress in the medium would produce a force of attraction between the
poles.[8]

In rendering the argument quantitative, as a series of mathematical
propositions, Maxwell employed the formalism of "resolution of
stresses," as presented, in particular, in Rankine's *A Manual of Applied
Mechanics* (1858). The result of Maxwell's quantitative calculation of the
state of stress in the vortex-filled medium can be represented, for
mnemonic convenience, in tensor form as follows:

$$\overset{\leftrightarrow}{\mathbf{T}} = \frac{\rho_m}{4}\, \boldsymbol{\omega}^*\boldsymbol{\omega}^* - \overset{\leftrightarrow}{\mathbf{I}}p_1 \tag{3.1}$$

where $\overset{\leftrightarrow}{\mathbf{T}}$ is the stress tensor, represented as a dyadic, which Maxwell denoted by components "p_{xx}, p_{yy}, and p_{zz} . . . the normal stresses . . . and . . . p_{yz}, p_{zx}, and p_{xy} . . . the tangential stresses"; ρ_m is the mass density of the fluid medium, where the subscript m has been appended to Maxwell's original notation for explicitness; and Maxwell's p_1 denotes an as-yet-arbitrary base-level "simple hydrostatic pressure," which appears in all three normal stresses, as indicated here by means of the unit dyadic $\overset{\leftrightarrow}{\mathbf{I}}$. Finally, $\boldsymbol{\omega}^*$ is a vector field characterizing the rotational motions of the vortices, which Maxwell denoted by components vl, vm, vn, where l, m, n are the "direction-cosines of the axes of the vortices," and "v is the linear velocity at the circumference of each vortex." (For the specific case of a vortex with uniform angular velocity $\boldsymbol{\omega}$ and circular cross section of radius a, we have $\boldsymbol{\omega}^* = a\boldsymbol{\omega}$; as we shall see, Maxwell was considering a somewhat more general case.[9])

In conjunction with his calculation of forces on the basis of the stress tensor $\overset{\leftrightarrow}{\mathbf{T}}$, Maxwell stipulated definite correspondences between the mechanical variables $\boldsymbol{\omega}^*$ and ρ_m and the electromagnetic variables they were supposed to represent. Thus, $\boldsymbol{\omega}^*$ was taken to correspond to a vector with components α, β, γ, which was identified as "the force which would act upon that end of a unit magnetic bar which points to the north," and which we shall denote **H**:

$$\boldsymbol{\omega}^* \text{ corresponds to } \mathbf{H} \tag{3.2}$$

Maxwell, in fact, expressed this as a strict equality, writing, in component form,

$$\alpha = vl, \quad \beta = vm, \quad \gamma = vn$$

that is, in our notation,

$$\mathbf{H} = \boldsymbol{\omega}^*$$

and thereafter dropping all reference to the vector $\boldsymbol{\omega}^*$ itself. This complete identification of the mechanical variable $\boldsymbol{\omega}^*$ with the electromagnetic variable **H**, of which it was the mechanical representation, was consonant with the theoretical and realistic, rather than analogical and heuristic, intent of the mechanical representation: $\boldsymbol{\omega}^*$ *is* **H** in the theory of molecular vortices. On the other hand, it will be convenient for the purposes of our analysis to maintain distinct mechanical and electromagnetic notations; thus, hereafter, we shall render Maxwell's α, β, γ sometimes as $\boldsymbol{\omega}^*$ and sometimes as **H**, depending on the specific context. A discussion of the theory in a letter from Maxwell to Thomson furnishes precedent for maintaining this kind of conceptual separation of mechanical and electromagnetic aspects, and we shall treat other variables in the theory in the same manner. (The choice of a specific rendering, in either electromagnetic or mechanical notation, will, of course, be a matter of interpretation, rather than of straightforward translation.[10])

One further correspondence between mechanical and electromagnetic variables was required for the treatment of magnetic forces in terms of the mechanical stresses in the system of molecular vortices – namely, the density of the fluid medium multiplied by π, $\pi\rho_m$, was taken to correspond to the electromagnetic variable μ, where "μ represents the magnetic inductive capacity of the medium at any point referred to air as a standard":

$$\pi\rho_m \text{ corresponds to } \mu \tag{3.3}$$

On the basis of correspondences (3.2) and (3.3), equation (3.1) was then transformed as follows:

$$\overleftrightarrow{\mathbf{T}} = \frac{\mu}{4\pi} \, \mathbf{HH} - \overleftrightarrow{\mathbf{I}} p_1 \tag{3.4}$$

Maxwell was able to complete his derivation of magnetic forces by evaluating the appropriate spatial derivatives of the components of normal and tangential stresses, again as specified in Rankine's *Applied Mechanics*. That amounted to taking the divergence of equation (3.4):

$$\mathbf{F} = \text{div } \overleftrightarrow{\mathbf{T}}$$

$$= \mathbf{H}\left(\frac{1}{4\pi} \text{ div } \mu\mathbf{H}\right) + \frac{\mu}{8\pi} \text{ grad } H^2 + \mu\mathbf{H} \times \frac{1}{4\pi} \text{ curl } \mathbf{H} - \text{ grad } p_1 \tag{3.5}$$

where $\mathbf{F}$ is the magnetic force per unit volume, which Maxwell denoted by components $X, Y, Z;$ div, **grad,** and **curl** are differential operators, which Maxwell here wrote in component form, but in later work characterized as quaternion operators; and $\times$ denotes the vector product.[11] Maxwell interpreted the first term on the right-hand side of equation (3.5) as giving the forces on magnetic poles, characterized as regions of nonzero divergence of the "magnetic induction" $\mu\mathbf{H}$; the question of the existence of nonzero values for div $\mu\mathbf{H}$ was finessed. The second term was understood as giving the force on "a paramagnetic or diamagnetic body placed in a field of varying magnetic force," characterized by nonzero values of the gradient of H^2. The third term was interpreted as giving the magnetic forces on electric currents, characterized as regions of non-zero **curl H**. Finally, the fourth term was given no electromagnetic interpretation at all, being characterized only as "the effect of simple pressure" in the medium.[12]

It was this successful quantitative calculation of magnetic forces – referred ultimately to the centrifugal forces associated with the rotations of the vortices – that was taken to fulfill the requirement of "accounting for" the "observed [magnetic] forces," thus establishing the basic legitimacy of the hypothesis of molecular vortices and justifying the correspondences between mechanical and electromagnetic variables stipulated in correspondences (3.2) and (3.3). Several caveats must, however, be raised in this connection. First, let us consider the final term in equation (3.5): Here – and elsewhere in the theory, as we shall see – Maxwell was unable to assign a definite electromagnetic interpretation to a mechanical effect predicted in the theory; the theory was in this respect not without loose ends. Second, it must be noted that certain approximations were associated with the calculations. We have written correspondence (3.3), between the magnetic inductive capacity μ and the density of the medium ρ_m, in the determinate form in which Maxwell carried it forward to Part III of the paper, and there used it in the crucial calculation of the velocity of wave propagation in the magnetoelectric medium.[13] This form, which is in accord with the mechanical stress tensor as given in equation (3.1), with the coefficient $\rho_m/4$, was ultimately based on Maxwell's initial calculations of the pressure differentials in the system of vortices, which had assumed that each vortex had circular cross section and uniform density as well as uniform angular velocity throughout. Maxwell had gone on to consider a more general case – all vortices having similar but arbitrary cross section and similar but arbitrary distributions of density and angular velocity – and had concluded, on the basis of very general scaling arguments, that the pressure differentials would be at most modified by a multiplicative constant, so that the coefficient $\rho_m/4$ in equation (3.1) would be replaced by $C\rho_m$ (where C is a number of order 1, and ρ_m is now an average density), and the change could be absorbed by having correspondence (3.3) become

$$4\pi C \rho_m \text{ corresponds to } \mu \qquad (3.6)$$

This result was quite general and was applicable to the more complicated vortices that Maxwell considered in later parts of the paper; the constant C was, however, not easy to calculate for those cases, and Maxwell in fact made use of correspondence (3.3) when he needed a definite result, thus making use of the simpler result as an approximation. The general argument had not been in vain, however, for it showed that the forms of equations (3.1) and (3.3) – and hence (3.5) – would still be valid for the more general case, so that the error made in approximating would amount at most to a misestimation of the value of a constant.[14]

That kind of approximation, involving possible misestimation of numerical factors – hopefully "not . . . mak[ing] much difference in the numerical result[s]"[15] and presumably involving no changes in the forms of the equations – was to be employed throughout the paper. There was a further element of arbitrariness involved in the stipulation of correspondences between mechanical and electromagnetic variables, as in correspondences (3.2) and (3.3). It will be noted that in the transformation of the stress tensor of equation (3.1) into that of equation (3.4), all that was required was that

$$\frac{\rho_m}{4}\, \omega^*\omega^* \text{ corresponds to } \frac{\mu}{4\pi}\, \mathbf{HH}$$

or, multiplying through by 4π,

$$\pi\rho_m\omega^*\omega^* \text{ corresponds to } \mu\mathbf{HH}$$

The stipulated correspondences (3.2) and (3.3) for μ and $\mathbf{H}$ individually do indeed guarantee this last correspondence for the combination of μ and $\mathbf{H}$ – linear in μ and quadratic in $\mathbf{H}$ – which enters into the stress tensor and hence into the terms of equation (3.5), but the following less restrictive stipulations for μ and $\mathbf{H}$ would also suffice:

$$\sqrt{b}\,\omega^* \text{ corresponds to } \mathbf{H} \tag{3.2'}$$

$$\frac{1}{b}\,\pi\rho_m \text{ corresponds to } \mu \tag{3.3'}$$

where b is a dimensionless parameter that can have any value greater than zero. The parameter b defines a one-parameter family of possible mechanical representations, of which Maxwell chose, as a matter of convenient convention, the one with $b = 1$. In fact, that convention had no effect on Maxwell's substantive conclusions; we have made the convention explicit and shall trace its propagation through the theory precisely in order to show that nothing really hangs on it. (In particular, we shall find that the calculations concerning waves in the medium in Part III of "Physical Lines" are not affected by the value of the parameter b – see Chapter 5.[16])

In summarizing the results of Part I of "Physical Lines," Maxwell observed that he had "shewn that all the forces acting between magnets, substances capable of magnetic induction, and electric currents, may be mechanically accounted for," on the basis of certain stresses in the "surrounding medium," with the "hypothesis of vortices [giving] a probable answer" to the question of the "mechanical cause" of these stresses.[17] This was the core of the theory of molecular vortices, both historically and rhetorically; what remained for Part II of "Physical Lines" was to further extend the theory, building out from the core established in Part I, in such a way as to maintain the coherence of the theory.

2   *Molecular vortices applied to electric currents: extending the theory*
        Part II of "Physical Lines" was published in two installments, in April and May of 1861, and was entitled "The Theory of Molecular Vortices Applied to Electric Currents." In that part of the paper, Maxwell undertook to extend the theory of molecular vortices to include electromagnetism (the production of magnetic fields and hence magnetic forces by electric currents) and electromagnetic induction (the production of electric currents by changing magnetic fields). The rhetorical structure involved starting with the simple model of Part I, dealing with magnetostatic forces, and then proceeding to enrich the model in Part II, so as to account for additional phenomena as well: The basic model involved only the molecular vortices themselves; the enriched model put monolayers of small spherical particles between the vortices, thus dividing the "magnetic medium" into "small portions or cells, the . . . cell walls being composed of . . . single strat[a] of [the] spherical particles." As the layers of small spherical particles between adjacent vortices did not appear in Thomson's original suggestions for a theory of molecular vortices in 1856, and as there was no hint of this elaboration of the theory in Maxwell's letters and discussions of 1857 and 1858, it would seem that Maxwell's rhetoric was again in conformity with the historical reality: The vortices themselves, and their relationship to magnetic forces, came first in Maxwell's thinking, and the small spherical particles were added later, in an effort to further refine and extend the theory.[18]

As particular motivation for putting in the small particles, Maxwell offered two kinds of considerations. First, there was the need to extend the range of the theory to embrace the connection between magnetic fields and electric currents: "We . . . now come to inquire into the physical connexion of these vortices with electric currents." (This "connexion" included, for Maxwell, both electromagnetism and electromagnetic induction.) The other motivation, according to Maxwell, was to remedy a mechanical defect in the system of molecular vortices as it had been envisioned in Part I of "Physical Lines." If one viewed the system of molecular vortices in the magnetoelectric medium as a real mechanical system – if one worried about its mechanics as would a mechanical engineer – one came up against the following problem: "I have found great difficulty," wrote Maxwell,

> in conceiving of the existence of vortices in a medium, side by side, revolving in the same direction about parallel axes. The contiguous portions of consecutive vortices must be moving in opposite directions; and it is difficult to understand how the motion of one part of the medium can coexist with, and even produce, an opposite motion of a part in contact with it.

Maxwell went on to observe that this kind of problem had been encountered, and solved, by the designers of mechanical devices:

> In mechanism, when two wheels are intended to revolve in the same direction, a wheel is placed between them so as to be in gear with both, and this wheel is called an "idle wheel." The hypothesis about the vortices which I have to suggest is that a layer of particles, acting as idle wheels, is interposed between each vortex and the next, so that each vortex has a tendency to make the neighbouring vortices revolve in the same direction with itself.

(Maxwell's pictorial rendering of the system of vortices and idle-wheel particles – as in Figure 2.2 – is redrawn in Figure 3.1, where the arrows indicate the simple case of all vortices rotating in the same sense and with the same rotational speed.[19])

In the rhetoric of Maxwell's presentation, the mechanical problem of coupling neighboring vortices was the proximate motivation for the introduction of the small particles acting as idle wheels. Subsequent investigation of the behavior of these small particles was to disclose that they might move under certain circumstances, and it was finally to become clear that the idle-wheel particles would move in such a way as to properly represent the electric current, with changes in their motion corresponding to electromagnetic induction. The means undertaken to resolve a mechanical problem in the system of molecular vortices thus ultimately – according to Maxwell's rhetoric – served to facilitate the extension of the theory to embrace additional phenomena. It would perhaps be over-credulous to conclude that this must represent the exact order in which the ideas occurred to Maxwell; what is important, rather, is that the mechani-

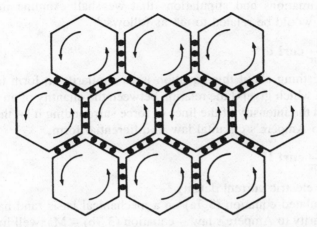

Figure 3.1. Vortices and idle-wheel particles. (Redrawn from Maxwell, "Physical Lines," Plate VIII, opposite p. 488; cf. Figure 2.2.)

cal difficulties with the model and the need to extend the range of the theory were accommodated by the single step of introducing the small particles, and this provided a strong sense of mutual reinforcement. The rhetorical order is, however, interesting for its own sake: Maxwell presented himself as mechanical engineer of the magnetoelectric medium, first looking to mechanical problems, and only later noticing the electromagnetic significance of the solutions to those problems.

Having thus, as mechanical engineer of the medium, introduced the small particles to function as idle wheels, Maxwell noticed that in a region of homogeneous magnetic field, where adjacent vortices would have equal ω^* and hence equal surface velocities, the particles making up the cell walls would behave as ordinary idle wheels, rotating but undergoing no spatial translation (Figure 3.2a); on the other hand, in a region of inhomogeneous magnetic field, adjacent vortices would have slightly different surface velocities, giving rise to translational motions of the idle-wheel particles (Figure 3.2b). (Maxwell was able to cite mechanical engineering precedent for this kind of situation: "In Siemens's governor for steam-engines, we find idle wheels whose centres are capable of motion"; today, one can point to ball bearings as an even more vivid analogue of Maxwell's layers of movable idle wheels.[20]) Defining a vector p, q, r – which we denote by ι – to be the net flux density of idle-wheel particles, averaged over a pseudodifferential volume containing many vortices, Maxwell calculated, on a purely kinematic basis (given certain approximations and stipulations that we shall examine in due course), that ι would be related to ω^* as follows:[21]

$$\iota = \frac{1}{4\pi} \text{ curl } \omega^* \tag{3.7a}$$

The interesting thing about this equation is its similarity in form to the "equation . . . which give[s] the relation between the quantity of an electric current and the intensity of the lines of force surrounding it" – that is, its similarity to Ampère's circuital law in differential form,

$$J = \frac{1}{4\pi} \text{ curl } H \tag{3.7b}$$

where J is the electric current density.[22]

Having calculated equation (3.7a) on a mechanical basis, and having noted its similarity to Ampère's law – equation (3.7b) – Maxwell immediately drew the following conclusion: "It appears therefore, that according to our hypothesis, an electric current is represented by the transference of the moveable particles interposed between the neighbouring vortices." The logic of this conclusion can be more explicitly represented as follows [where the designations i and f – as in (3.7i) and (3.7f) – refer to the initial and final correspondences, respectively]:[23]

Given the already established correspondence

$$\omega^* \text{ corresponds to } H \tag{3.7i}$$

and given the parallel mechanical
and electromagnetic relationships

$$\iota = \frac{1}{4\pi} \text{ curl } \omega^* \tag{3.7a}$$

$$J = \frac{1}{4\pi} \text{ curl } H \tag{3.7b}$$

it is concluded that

$$\iota \text{ corresponds to } J \tag{3.7f}$$

(3.7)

With this, Maxwell had extended the theory of molecular vortices in two ways: He had incorporated a new electrical variable into the theory – namely, the electric current J – by identifying it with the flux of idle-wheel particles ι, and he had incorporated a new electromagnetic relationship into the theory – namely, Ampère's law – by identifying it with the relationship between the rotations of the vortices and the motions of the idle-wheel particles.

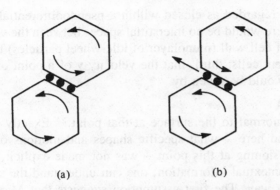

Figure 3.2. Adjacent vortices: (a) with equal surface velocities; (b) with unequal surface velocities.

The rhetorical strategy exemplified in schema (3.7) performed two basic functions for Maxwell. First, it explicitly guaranteed that the new variable and the new relationship were properly linked to the original core of the theory – in parallel at the mechanical and electromagnetic levels – so that the coherence and consistency of the theory were strictly maintained even as it was enlarged in scope; if additional variables and relationships could be introduced in a similar way, the result would be a comprehensive theory that was guaranteed, explicitly and by formal structure, to be coherent and consistent throughout. Second, by presenting correspondence (3.7f) as a discovery, as a final step, Maxwell was arguing that this extension of the theory was natural: A reasonable mechanical improvement had been introduced into the model – namely, the idle-wheel particles – and the result had been, according to Maxwell's rhetoric, a serendipitous extension of the power of the theory, yielding an understanding of the physical nature of the electric current and its mode of connection with the magnetic field. Maxwell was in this way – by using this rhetorical scheme – arguing that the modification of the model through the introduction of the idle-wheel particles was not ad hoc to the extension of the theory, but rather followed in a natural way from the core assumptions of the theory. Thus, the rhetoric of schema (3.7) was intended to demonstrate that the attendant extension of the theory was both coherent and natural.

The success of schema (3.7) in ensuring the coherence of the theory, however, was in turn dependent on the coherence and exactness of the mechanical calculations – as of equation (3.7a) – that underwrote it. In his calculation of equation (3.7a), Maxwell had made three assumptions concerning the vortices and idle wheels: first, that the surface of each vortex cell could be regarded as closed within a pseudodifferential volume; second, that there would be no interstitial spaces between the vortex cells, each portion of cell wall (monolayer of idle-wheel particles) being shared by two adjacent cells; third, that the velocity **v** of a point on the surface of a vortex would be given by

$$\mathbf{v} = \boldsymbol{\omega}^* \times \mathbf{n} \tag{3.8}$$

where **n** is the unit normal to the surface at that point.[24] Exactly what Maxwell had in mind here – what specific shapes and motions of the vortices he was envisioning at this point – was not made explicit, but, relying in part on contextual information, one can understand the situation in the following way: The first assumption suggests that Maxwell was no longer regarding the vortex cells as filamentous with macroscopic length, but rather was considering them as compact in all dimensions; in Part III of "Physical Lines," the vortex cells would be explicitly characterized as close to spherical in shape – approximable to spheres – and it would appear that Maxwell was already regarding the vortex cells in that way in Part II.[25] [As discussed in connection with correspondence (3.6), a general scaling argument guaranteed that the form of the magnetic stress tensor – and hence the magnetic forces – would not be altered by this change in shape of the vortex cells; the multiplicative constant involved in that correspondence could then be treated by approximation.] The basis for Maxwell's second assumption is clarified by the figure that he supplied (reproduced as Figure 2.2, and redrawn as Figure 3.1), in which the vortices were pictured, in cross section, as a plane-filling array of hexagons;[26] correspondingly, in three dimensions one envisions a space-filling array of polyhedra – perhaps the (irregular) dodecahedra specified by the hexagonal close packing of spheres, which can be approximated to spheres for many calculations.

Maxwell's third assumption, as represented in equation (3.8), can be understood as follows: Assume that each vortex has uniform angular velocity throughout[27] and is being approximated as spherical in shape. Taking **ω** to be the angular velocity, a the radius of the sphere, **r** the radius vector to a point on the surface, and **n** the unit normal to the surface at that point, then the velocity **v** of that point will be given by

$$\mathbf{v} = \boldsymbol{\omega} \times \mathbf{r} \tag{3.8a}$$
$$= \boldsymbol{\omega} \times a\mathbf{n} \tag{3.8b}$$
$$= a\boldsymbol{\omega} \times \mathbf{n} \tag{3.8c}$$
$$= \boldsymbol{\omega}^* \times \mathbf{n} \tag{3.8d}$$

exactly as in equation (3.8).

The final ingredient in the calculation of equation (3.7a) was the stipulation that the surface densities σ of the layers of idle-wheel particles – measured in electromagnetic units of charge per unit area, rather than by number of particles – be given by[28]

$$\sigma = \frac{1}{2\pi} \qquad (3.9)$$

That gave the proper factor of $1/4\pi$ in equation (3.7a), thus justifying the precise correspondence (3.7f). The specific numerical value in equation (3.9) may be seen as the direct result of Maxwell's stipulation of the value 1 for the parameter b (as discussed in Section 1 of this chapter); an undetermined b would have been reflected in equation (3.9) as follows:

$$\sigma = \frac{\sqrt{b}}{2\pi} \qquad (3.9')$$

Maxwell's initial convention, $b = 1$, was thus propagated through the value of σ.[29] (The consequences of this, as concerns Maxwell's subsequent calculation of the velocity of waves in the magnetoelectric medium, will be explored in Chapter 5.) In sum, equation (3.7a) may be seen as the result of credible although approximative mechanical assumptions and stipulations.[30] The mechanical calculation of equation (3.7a) then served to underwrite Maxwell's use of schema (3.7), yielding a coherent extension of the theory to embrace a mechanical representation of the electric current in terms of "the transference of the moveable particles interposed between the neighbouring vortices."[31]

The second major task of Part II of "Physical Lines" was the extension of the theory to include electromagnetic induction, and the mechanical considerations involved in this pertained to the circumstances accompanying changes in the rotational motions of the vortices. Such changes would be engendered by tangential forces exerted by the idle-wheel particles on the surfaces of the vortices, and these would be accompanied by inertial reaction forces exerted by the vortices on the idle-wheel particles. These reaction forces were characterized by a vector field with components P, Q, R – here to be denoted τ_1 – such that the component of τ_1 tangential to the surface of a vortex at any point, multiplied by $\sigma/2$ (i.e., $1/4\pi$), would give the surface density, at that point, of the tangential force exerted by the vortex surface on the adjacent idle-wheel particles.[32] (Following Maxwell, we shall defer for a time discussion of the physical problems associated with the exertion of a tangential force by a fluid surface.) Invoking the same assumptions and stipulations that had been used in the calculation of equation (3.7a), and requiring that the work done by the forces $\boldsymbol{\tau}_1$, integrated over the surface of a vortex, be equal to the change in rotational kinetic energy of that vortex, Maxwell calculated that[33]

$$-\mathbf{curl}\ \boldsymbol{\tau}_1\ =\ \pi\rho_m\frac{\partial\omega^*}{\partial t} \qquad (3.10a)$$

The interesting thing about this equation was its similarity – indeed identity, in Maxwell's dual-purpose notation – with "the relation between changes in the state of the magnetic field and the electromotive forces thereby brought into play" – that is, Faraday's law of electromagnetic induction in differential form,

$$-\mathbf{curl}\ \mathbf{E}_1\ =\ \mu\frac{\partial\mathbf{H}}{\partial t} \qquad (3.10b)$$

where $\mathbf{E}_1$ refers to "the electromotive forces . . . brought into play" by "changes in the . . . magnetic field." (Maxwell went on to show that this is the part of the electric field that is derivable from a vector potential.[34])

Having calculated equation (3.10a) on a mechanical basis, and having noted its similarity to Faraday's law of induction – our (3.10b) – Maxwell was able to conclude that "the forces exerted on the layers of particles between the vortices" represent, "in the language of our hypothesis, . . . electromotive forces." Once again, a schematic representation will help to render explicit the logical and rhetorical foundation for this conclusion:

Given the already established correspondences

ω^* corresponds to $\mathbf{H}$

$\pi\rho_m$ corresponds to μ (3.10i)

and given the parallel mechanical and electromagnetic relationships

$$-\mathbf{curl}\ \boldsymbol{\tau}_1\ =\ \pi\rho_m\frac{\partial\omega^*}{\partial t} \qquad (3.10a)$$

$$-\mathbf{curl}\ \mathbf{E}_1\ =\ \mu\frac{\partial\mathbf{H}}{\partial t} \qquad (3.10b)$$

it is concluded that

$\boldsymbol{\tau}_1$ corresponds to $\mathbf{E}_1$ (3.10f)

(3.10)

With that, Maxwell had once again extended the theory of molecular vortices in two ways: He had incorporated yet another electromagnetic variable into the theory – the electromotive force $\mathbf{E}_1$ – by identifying it with the forces exerted by the surfaces of the vortices on the idle-wheel particles, and he had incorporated yet another electromagnetic relationship into the theory – namely, Faraday's law of electromagnetic induction – by identifying it with the relationship between torques exerted on the vortices and their rotational accelerations.[35]

The reasoning summarized in schema (3.10) once again – just as with schema (3.7) – served to demonstrate that the coherence of the theory was being maintained even as its scope was being enlarged; that was guaranteed by the explicit parallel linkages of variables at the mechanical and electromagnetic levels. Also, the further extension of the theory was once again exhibited as a happy consequence of reasonable mechanical improvements to the model – to wit, the insertion of the idle-wheel particles. The cogency of the argument again rested ultimately on the mechanical calculation involved [leading in this instance to equation (3.10a)], and the assumptions and stipulations involved were, for the most part, the same basically reasonable although approximative assumptions and stipulations that had been used before in the calculation of equation (3.7a). For the most part, then, the strategy for extending the theory embodied in schemata (3.7) and (3.10) had been applied in a consistent manner. These extensions of the theory, however, gave rise to some new difficulties, which were not easily resolved.

3   *Difficulties*
       The molecular-vortex model, as enriched by the addition of the idle-wheel particles, was subject to certain physical and mechanical problems for which Maxwell evidently did not yet have good solutions when he was preparing Part II of "Physical Lines" for publication (in two installments, in April and May of 1861); ultimately, however, those problems were, in their turn, to bring forth further refinements to the model, which provided the background for the extension of the theory to electrostatics, the introduction of the displacement current, and the formulation of the first version of the electromagnetic theory of light – all in Part III of "Physical Lines," which was published in January of 1862, after an eight-month hiatus. As the eight-month delay suggests – and as further evidence and analysis will support (see also Chapter 2, Section 3) – the progression from mechanical problems in the model, as developed in Part II, to solutions leading to new extensions of the theory, in Part III, was no mere rhetorical device, but instead reflected historical circumstance: Maxwell did indeed encounter severe and fundamental problems in the

development of the theory, and it took him the better part of a year to resolve them.

The most glaring of the problems that bedeviled Maxwell in Part II of "Physical Lines" – as reflected especially in the April installment – was the artificiality of the idle-wheel hypothesis, considered as a physical hypothesis. Considered as an engineering device, the system of idle-wheel particles represented quite a natural solution to the problem of coupling the motions of adjacent vortices rotating in the same sense. As a physical hypothesis, however – as an attempt to get at a "true interpretation of the phenomena" – Maxwell had had to admit that the hypothesis of the idle-wheel particles was a "somewhat awkward" hypothesis of "provisional and temporary character," which he did "not bring . . . forward as . . . existing in nature."[36] Maxwell's doubts concerning the idle-wheel hypothesis are discussed more fully in Chapter 2; some understanding of why, nevertheless, Maxwell persisted for a time in making use of this hypothesis is furnished by consideration of its logical and rhetorical role in connection with schemata (3.7) and (3.10). The hypothesis of the small particles, by its simultaneous service both to improve the mechanics of the model and to facilitate coherent extension of the theory, thereby established itself as central to the theory, even though Maxwell had serious reservations concerning it. One thinks of the testimony of novelists to the effect that characters and situations sometimes seem to take control of the novel, determining their own development by an inner logic, and thereby generating outcomes that the writer had neither foreseen nor endorsed at the outset. In a similar fashion, it would appear that the developmental process by which the theory of molecular vortices was being elaborated acquired a momentum and a logic of its own, leading Maxwell to elaborations of the theory with which he was not fully comfortable.[37]

Another problem connected with the idle-wheel particles concerned their interaction with the surfaces of the vortices: How could a fluid surface exert tangential forces on the idle-wheel particles? Maxwell discussed this problem explicitly in his summary and conclusion at the end of Part II of "Physical Lines," calling attention also to another aspect of the problem: It had been assumed for purposes of some of the calculations that the angular velocity would be uniform throughout a given vortex, and it had been assumed throughout that the distribution of angular velocity would at least be similar for all the vortices. In order, however, for these conditions to be maintained in a situation where the rotational motions of the vortices were changing – in response to forces exerted at their surfaces – it would be necessary that the motions of the interior strata of the vortices be coupled somehow to the motions of the exterior strata and the surface, so that the changes in angular velocity would be properly transmitted to the interior regions of each vortex. If the fluid medium were frictionless, neighboring strata of a vortex would slip against each other without interacting, and changes in motion would not be transmitted to the interior; if the fluid were viscous, the motions of neighboring strata would be coupled by viscous shear stresses – tangential forces – but the action of these would be accompanied by conversion of some of the rotational energy of the vortex into heat, and such frictional losses of magnetic field energy were unacceptable in a realistic theory.[38]

With hindsight, we know that Maxwell was able to solve this problem quite neatly – in Part III of "Physical Lines," after the eight-month hiatus – by endowing the material within the vortex cells with elastic properties (elastic materials can sustain nondissipative shear stresses). When he was writing the closing pages of Part II of the paper, however, he apparently did not yet see his way clearly – if at all – to the further modifications and extensions of the theory that were ultimately to appear in Parts III and IV; his solution to the problem in the closing passages of Part II was thus strictly by fiat: "We must therefore conceive that . . . the interior strata of each vortex receive their proper velocities from the exterior strata without slipping, that is, the angular velocity must be the same throughout each vortex."[39] Maxwell did not elaborate on how one was to "conceive" this behavior for a fluid vortex; the behavior described seems rather to be that of a rigid solid body. [In fact, if equations (3.8a) – (3.8d) are to hold true, even when the angular velocity is changing, then the behavior described is exactly that of a rigid sphere.] With hindsight, once again, it is easy enough to see that endowing the vortex material with elastic properties represents a compromise of sorts – as between fluidity and rigidity – and thus promises to resolve the problem. At the time, however, the resolution was not quite so straightforward, and Maxwell clearly had not yet seen his way through the problem when he published Part II of "Physical Lines." [Another hindrance to seeing elasticity as the solution was that Thomson's "dynamical" approach entailed an attempt to see elasticity as resulting from motion in the mechanical substratum, rather than from primitive static forces between the molecules (see Chapter 2, Section 2); that would have militated against a straightforward assumption of elasticity for the vortex material.]

Another problem that faced Maxwell, one for which he evidently had no solution in mind when he published Part II of "Physical Lines," was that of the extension of the theory to embrace electrostatics. Maxwell clearly wanted to set up his own theory as a rival to Weber's action-at-a-distance alternative, and to do that in a serious way, he needed to extend the theory of molecular vortices to include electrostatics, in order to match Weber's comprehensiveness of coverage. Part II of "Physical Lines," however, closed with what was clearly intended to be a final summary and conclusion; no further installments were contemplated, and no extension to electrostatics was promised.[40] In the event, Maxwell did finally find a way to extend the theory to electrostatics, but that came only after the eight-month delay.

The nature of the difficulty that delayed Maxwell becomes clear if we ask how the logic of theory extension exhibited in schemata (3.7) and (3.10) would have applied in the extension of the theory to electrostatics. The natural route to electrostatics through this kind of formalism was in fact adumbrated at one point in Part II, where Maxwell suggested that the idle-wheel "particles . . . in our hypothesis represent electricity."[41] Given that the flux density of idle-wheel particles was identified as the electric current, it was only natural to identify the particles themselves as "electricity," with the implication that an excess accumulation of these particles in some part of the medium (over and above the normal amount associated with the monolayers of surface density $\sigma = 1/2\pi$) could be identified as a static electric charge. Represented in terms of the usual logical schema, the reasoning would be as follows:

Given the already established correspondence

$\qquad \iota$ corresponds to $\mathbf{J}$ $\qquad\qquad\qquad$ (3.11i)

and given the parallel mechanical
and electromagnetic relationships

$$\text{div } \iota + \frac{\partial \rho_p}{\partial t} = 0 \qquad\qquad\qquad (3.11a)$$

$$\text{div } \mathbf{J} + \frac{\partial \rho}{\partial t} = 0 \qquad\qquad\qquad (3.11b)$$

it is concluded that

$\qquad \rho_p$ corresponds to ρ $\qquad\qquad\qquad$ (3.11f)

(3.11)

where ρ_p is the excess density of idle-wheel particles, with equation (3.11a) following immediately from the assumption – implicit all along – that the idle-wheel particles are neither created nor destroyed; ρ is the electric charge density; and equation (3.11b) is what Maxwell called "the equation of continuity," expressing the conservation of electric charge.[42]

What made all of this less than straightforward was the circumstance that schema (3.11) was not consistent with schema (3.7). Nonzero values for the terms in equations (3.11a) and (3.11b) are inconsistent with equations (3.7a) and (3.7b), which are applicable only to closed circuits, and do not allow for the accumulation of charge:

$$\text{div } \mathbf{J} = \frac{1}{4\pi} \text{ div}(\text{curl } \mathbf{H}) \equiv 0$$

$$\frac{\partial \rho}{\partial t} = -\text{div } \mathbf{J} \equiv 0$$

$$\text{div } \iota = \frac{1}{4\pi}\text{div}(\text{curl } \omega^*) \equiv 0$$

$$\frac{\partial \rho_p}{\partial t} = -\text{div } \iota \equiv 0$$

Maxwell had long been aware that equation (3.7b), Ampère's circuital law in differential form, had restricted applicability, and that was a matter of concern to him;[43] having built the law into his mechanical theory in the form of equation (3.7a), he then had a theory of restricted applicability, which could not be extended to electrostatics in a straightforward way. I would suggest that this circumstance, along with the problem of the artificiality of the idle-wheel particles and the problem of shear stresses within the vortex medium, accounts for the air of pessimism and finality that characterized the conclusion to Part II of "Physical Lines." An eight-month hiatus in publication intervened after the appearance of Part II in April and May of 1861, but Maxwell did in fact continue to work on the theory – there is evidence that he made substantial progress while at home in Scotland in the summer of 1861 – and two further installments, Parts III and IV, were published early in 1862.[44]

4   *Molecular vortices and electrostatics*

Part III of "Physical Lines," published in January 1862, was entitled "The Theory of Molecular Vortices Applied to Statical Electricity." Not only did Maxwell extend the theory to embrace electrostatics; he also presented in this publication the initial forms of his two crucial innovations in electromagnetic theory – namely, the displacement current and the electromagnetic theory of light. Our concern here will be primarily with the extension to electrostatics; full discussions of the displacement current and the electromagnetic theory of light will be deferred to Chapters 4 and 5, respectively.

As had been the case in Part II of "Physical Lines," Maxwell motivated the further extension of the theory in Part III in two ways. On the one hand, there was the need to extend the range of the theory to embrace electrostatics:

> If we can now explain the condition of a body with respect to the surrounding medium when it is said to be "charged" with electricity, and account for the forces acting between electrified bodies, we shall have established a connexion between all the principal phenomena of electrical science.

The other motivation was – just as in Part II – the amelioration of a mechanical problem in the model; in Part III, the problem was to make sense – in mechanical terms – of the "tangential action" between the small particles and "the substance in the cells":

> I have not [in Part II] attempted to explain this tangential action, but it is necessary to suppose, in order to account for the transmission of rotation from the exterior to the interior parts of each cell, that the substance in the cells possesses elasticity of figure, similar in kind, though different in degree, to that observed in solid bodies.

The assignment of elastic properties to the substance of the cells was in fact to fulfill both needs, making possible the extension of the theory to electrostatics, while at the same time addressing the mechanical problem of the "tangential action."[45]

Once again, in the rhetoric of Maxwell's presentation, it was the mechanical problem that directly motivated the introduction of elasticity, and only later did it become clear that certain stresses in the elastic medium would properly represent the electrostatic field; once again, it would perhaps be overcredulous to believe, in the absence of any corroborating evidence, that this represents the actual order in which the ideas occurred to Maxwell. What is important is that the mechanical difficulties with the model and the need to extend the theory to electrostatics were both accommodated by the single step of endowing the vortex material with elasticity, and that provided mutual reinforcement. (There was additional motivation for introducing elasticity: "The undulatory theory of light requires us to admit this kind of elasticity in the luminiferous medium, in order to account for transverse vibrations. We need not be surprised if the magneto-electric medium possesses the same property."[46] This looked forward to the identification of the magnetoelectric and luminiferous media, which will be discussed in detail in Chapter 5.)

The magnetoelectric medium was thus to be pictured as follows: As before, the medium was to be viewed as cellular, with the "cell-walls . . . composed of [monolayers of] particles which are very small compared with the cells." As before, the cells were taken to be nearly spherical, space-filling polyhedra; Maxwell explicitly pointed out that he was proceeding "on the supposition that the cells are spherical. The actual form of the cells . . . does not differ from that of a sphere sufficiently to make much difference in the numerical [calculations]."[47] As before, the material within a given cell might rotate about some axis, and that rotation would represent a magnetic field. At this point in the development of the model, however, the material within a cell was to be envisioned as a single chunk or blob of elastic-solid material, rather than a parcel of fluid. If one of these pseudospherical elastic blobs were to rotate about a given axis, it would tend to bulge equatorially and flatten at its poles; a string of rotating blobs corresponding to a magnetic field line would therefore tend to contract along its length as a result of the centrifugal forces, just like the vortex filaments representing field lines in the simpler mechanical representation with which Maxwell had begun in Part I of "Physical Lines." Thus, magnetic forces similar to those calculated in Part I would still be produced, and it was apparently hoped – though not explicitly argued – that the stress tensor calculated in Part I would still constitute a good approximation.[48] The mechanical account of electromagnetic induction, as presented in Part II, was preserved, and indeed enhanced. The problem of the transmission of changes of rotational velocity throughout the material of a given vortex had been solved by endowing the vortex material with elastic properties. Also, it was now easier to picture the mechanical coupling between the idle-wheel particles and the vortices: The idle-wheel particles would roll without slipping upon, and exert tangential forces upon, the solid surfaces of the elastic blobs with which they were in contact.

The introduction of elasticity facilitated the extension of the theory of molecular vortices to electrostatics by showing the way to a modification of equation (3.7a), the mechanical analogue of Ampère's law. (The larger context for the modification of Ampère's law will be taken up in Chapter 4.) The mechanical equation had followed from the assumption that each vortex, approximated as spherical in shape, had uniform angular velocity throughout, and thus rotated as a rigid sphere. If, however, the molecular vortices were to be regarded as rotating *elastic* spheres, then, as Maxwell put it, the equation would have to be "correct[ed] . . . . for the effect due to the elasticity of the medium." That was accomplished by inserting a correction term in the equation:

$$\iota = \frac{1}{4\pi} \text{ curl } \omega^* + \iota_{\text{elas def}} \qquad (3.12)$$

where $\iota_{\text{elas def}}$ represents the flux of idle-wheel particles attributable to progressive elastic deformation of the elastic vortex spheres. These deformations would, in turn, entail elastic reaction forces, exerted by the vortex spheres on the idle-wheel particles; characterizing the reaction forces by a vector field τ_2, and denoting the shear modulus of the elastic medium as m, Maxwell calculated the following correction term:

$$\iota_{\text{elas def}} = -\frac{1}{4\pi^2 m} \frac{\partial \tau_2}{\partial t}$$

The final form for the equation giving the flux density of idle-wheel particles ι was[49]

$$\iota = \frac{1}{4\pi} \left(\text{curl } \omega^* - \frac{1}{\pi m} \frac{\partial \tau_2}{\partial t} \right) \qquad (3.13)$$

The significance of equation (3.13) for the extension of the theory of molecular vortices to electrostatics lay in the fact that the divergence of ι, the flux of idle-wheel particles, could now take on nonzero values, depending on the spatial pattern of the elastic forces τ_2:

$$\text{div } \iota = -\frac{1}{4\pi^2 m} \frac{\partial}{\partial t} \text{ div } \tau_2 \qquad (3.14)$$

Schema (3.11) could now be invoked (the logic was implicit rather than explicit at this point), with the conclusion that the excess density of idle-wheel particles ρ_p would correspond to the electric charge density ρ. Furthermore, Maxwell was able to calculate, from equations (3.14) and (3.11b), that ρ_p would be related to the pattern of elastic stresses τ_2 as follows:

$$\rho_p = \frac{1}{4\pi(\pi m)} \text{ div } \tau_2 \qquad (3.15a)$$

where m again is the shear modulus of the medium. This equation is similar in form to the electromagnetic equation relating the electric charge density ρ to the electrostatic field $\mathbf{E}_2$:

$$\rho = \frac{1}{4\pi c^2} \, \text{div } \mathbf{E}_2 \qquad\qquad (3.15b)$$

where c is an electromagnetic constant that Maxwell would soon identify explicitly as the ratio of units. Given the established correspondence between ρ_p and ρ, the conclusion could then be drawn that the elastic stresses τ_2 would correspond to the electrostatic field $\mathbf{E}_2$, with an associated correspondence between the shear modulus of the medium m and the electromagnetic constant c:[50]

Given the already established correspondence

ρ_p corresponds to ρ $\qquad\qquad$ (3.15i)

and given the parallel mechanical and
electromagnetic relationships

$$\rho_p = \frac{1}{4\pi(\pi m)} \, \text{div } \tau_2 \qquad\qquad (3.15a)$$

$$\rho = \frac{1}{4\pi c^2} \, \text{div } \mathbf{E}_2 \qquad\qquad (3.15b)$$

(3.15)

it is concluded that

τ_2 corresponds to $\mathbf{E}_2$

πm corresponds to c^2 $\qquad\qquad$ (3.15f)

There remained one final step in order to complete the extension of the theory of molecular vortices to electrostatics. As will be remembered (Chapter 2), in proposing what he was prepared to call a "physical theory," Maxwell had committed himself to developing a theory that not only would exhibit mechanical parallels for all known relationships among electromagnetic variables but also would give a mechanical account of the forces exerted in electromagnetic situations – that is, the magnetic forces exerted on poles, currents, and magnetic materials, and the electrical forces exerted on electrically charged bodies. That had been accomplished for magnetic forces in Part I of "Physical Lines," and Maxwell was now obliged to do the same for electrical forces; in particular, Maxwell set out to demonstrate that two electrically charged bodies, having charges η_1 and η_2 in electrostatic units, would exert forces of magnitude F on each other, F being given by

$$F = \frac{\eta_1 \eta_2}{r^2} \qquad\qquad (3.16)$$

where r is the distance between the two bodies. In doing this, Maxwell was also to identify the constant c as the ratio between electromagnetic and electrostatic units.

Maxwell derived the electrostatic force law, in the theory of molecular vortices, on the basis of energetic considerations. The energy U stored in the medium as a result of the elastic deformations of the vortex blobs was expressed as the space integral of an inner product of vectors representing respectively displacements and restoring forces:

$$U = - \frac{1}{2} \int \boldsymbol{\tau}_2 \cdot \boldsymbol{\delta} \, dV \qquad (3.17)$$

where $\boldsymbol{\tau}_2$ represents the restoring forces, $\boldsymbol{\delta}$ represents displacements, and the factor $1/2$ is included because the vector $\boldsymbol{\tau}_2$ represents the force exerted by *two* adjacent vortex surfaces on the layer of idle-wheel particles between. Given a linear relationship between $\boldsymbol{\tau}_2$ and $\boldsymbol{\delta}$, with the constant of proportionality depending on the shear modulus of the medium m, Maxwell arrived at the following relationship:

$$U = \frac{1}{8\pi^2 m} \int |\boldsymbol{\tau}_2|^2 \, dV \qquad (3.18)$$

Such an energy expression, involving the square of the field quantity $\boldsymbol{\tau}_2$, or the corresponding $\mathbf{E}_2$, leads in a straightforward way to an inverse-square force law, and Maxwell was able to deduce that bodies having charges e_1 and e_2 in electromagnetic units (these having been used all along in the paper) would exert forces of magnitude F on each other:

$$
\begin{aligned}
F &= \pi m \, \frac{e_1 e_2}{r^2} \\
&= c^2 \, \frac{e_1 e_2}{r^2}
\end{aligned}
\qquad (3.19)
$$

Finally, Maxwell observed that if this equation was to be assimilable to equation (3.16), it would have to be the case that

$$
\begin{aligned}
\eta &= ce \\
c &= \frac{\eta}{e}
\end{aligned}
\qquad (3.20)
$$

Thus, the constant c must be taken to be the ratio between electromagnetic and electrostatic units, "the number by which the electrodynamic measure of any quantity of electricity must be multiplied to obtain its electrostatic measure."[51]

That derivation was perhaps a bit less than Maxwell aspired to, as may be seen by comparing it with the derivation of magnetic forces in Part I of "Physical Lines." There, Maxwell had been able to show in detail how the rotations of the vortices would give rise to stresses in the medium; that enabled him to calculate a stress tensor in the medium, from which he then derived the observed forces. In particular, the tendency of the magnetic lines of force to move away from each other laterally and contract along their lengths was directly explained in terms of the correspondence with vortex filaments. In Part III, on the other hand, in the case of electric lines of force, there was no calculation of a macroscopic stress tensor in the medium, nor was there provided even a qualitative picture of the effect of the elastic forces in producing the behavior of electric lines of force. Instead, Maxwell presented an energetic derivation, which indeed sufficed for calculation of the resultant electrical force, but circumvented the problem of the specification of the precise mechanism of action of that force. Thus, although Maxwell had in a real sense satisfied the criterion that his theory account for both magnetic and electrical forces, he had not done full justice to the electrical case. Years later, in the *Treatise on Electricity and Magnetism*, Maxwell had to admit that, even then, a full solution to the problem eluded him.[52]

Conclusion

The formal strategy that Maxwell had employed to extend the theory of molecular vortices to embrace all of the basic electromagnetic phenomena – summarized in schemata (3.7), (3.10), (3.11), and (3.15) – had as its goal, above all, the formulation of a theory that would be coherent and self-consistent throughout its range of applicability: The effort had been undertaken in the Cambridge tradition of hypothetico-deductive deep theory, rather than the Scottish analogical tradition; thus, the mechanical images in the theory were intended to be taken not as disjunct heuristic pictures but rather as adding up to one consistent and realistically intended theory of electromagnetic phenomena. Aside from methodological concerns, there were also the requirements of electromagnetic theory to be considered. The basic task of the electrical theoretician at midcentury, as perceived within both the Continental and the British traditions, was to unify electromagnetic theory, accommodating the classical electrical and magnetic phenomena as well as the new, nineteenth-century discoveries by Oersted, Ampère, and Faraday within one coherent theory. Wilhelm Weber had managed to do that, most elegantly, within the Continental tradition; Maxwell perceived his task as that of accomplishing a parallel unification within the British tradition, so as to demonstrate that the field approach constituted a viable alternative to the action-at-a-distance approach.

Not without difficulty and delay, but nevertheless with ultimate success, Maxwell had managed, by the time he published Part III of "Physical Lines," to accomplish the unification of electromagnetic theory within a field-theoretic framework. Moreover, his pursuit of a unified account of electromagnetic phenomena had set the stage for the appearance of two theoretical novelties – the displacement current and the electromagnetic theory of light. The fundamental step toward the displacement current had been taken in the context of Maxwell's attempt to link the equations pertaining to electric current with the equations of electrostatics, through the equation of continuity; that had necessitated the modification of Ampère's law, which Maxwell accomplished at the mechanical level through the introduction of a new term expressing the effect of the elasticity of the vortices. There was, of course, a broader context for the modification of Ampère's law, and the electromagnetic interpretation of the modified equation was to emerge within that broader context (this will be explored further in Chapter 4). The crucial step toward the electromagnetic theory of light had been taken with the assignment of elastic properties to the magnetoelectric medium, but that was not the whole story. It was the chain of parallel linkages at the mechanical and electromagnetic levels – as summarized especially in schemata (3.7), (3.11), and (3.15) – that provided for a connection, in the theory, between magnetic forces and electrostatic forces, thereby allowing for the incorporation of the ratio of electrical units into the theory. As will be discussed in Chapter 5, that provided a basis for the identification of the ratio of units with the velocity of propagation of elastic waves in the magnetoelectric medium, leading to the initial form of the electromagnetic theory of light.

4

The introduction of the displacement current

The immediate context for Maxwell's initial modification of Ampère's law (Ampère's circuital law in differential form), through the introduction of a new term to be known as the "displacement current,"[1] was, as we have seen, his work on the theory of molecular vortices: His proximate aim in modifying Ampère's law was to extend the theory of molecular vortices to electrostatics, and his explicit interpretation at that point of the modified equation was as a mechanical calculation in the theory of molecular vortices, with the new term expressing the flux of the small idle-wheel particles owing to progressive elastic deformation of the vortices. All of the principal symbols and equations in "Physical Lines," however, had dual significance – mechanical and electromagnetic – and the modified Ampère's law, in its electromagnetic character, had broader connections and significance, transcending its proximate matrix in the theory of molecular vortices. That broader context must be taken into account if we are to achieve a full understanding of the origin of the displacement current and its significance in the history of electromagnetic theory.

The question of the origin of the displacement current has been, and continues to be, the object of much interest and concern: Each year many thousands of students in physics courses throughout the world learn that Maxwell, on the basis of theoretical considerations, modified Ampère's law, through the introduction of a new term called the displacement current, and thereby perfected the enduring foundation for modern electromagnetic theory. The centrality of this episode in the history of physics, its paradigmatic status as an example of theoretically motivated innovation, and its prominence in the pedagogy of physics have all contributed to making it a topic of prime concern for historians of physics. Unfortunately, however, the considerable effort hitherto invested in historical research on this subject has not been handsomely repaid, and the matter remains – as is generally acknowledged – obscure.

In a recent review of the Maxwell literature,[2] it was suggested that the root of the problem is that Maxwell scholars (at least some of them) have become mesmerized by certain details – such as the algebraic signs in Maxwell's various renderings of the equations relating to the displacement current – whereas what is needed instead for real progress in Maxwell studies is more attention to broader themes. One can hardly quarrel with the call for attention to broader themes, extending to "Maxwell's work as a whole [and] its British context," and this study takes that call seriously; I think one must take exception, however, to the judgment that "too much attention has been directed toward the signs in the equation[s]."[3] If one thing has been made clear by the cumulative record of historical investigation, it is that the history of the displacement current in particular, and Maxwell's electromagnetic theory in general, presents difficult problems of interpretation, and one does not make progress on difficult problems by neglecting important data. In particular, if one is concerned with the history of mathematical physics, one must not neglect the equations, for they are central. Indeed, as I shall argue in what follows, in the present case the details of the equations – including such matters as algebraic signs and grouping of terms – illuminate, and are reciprocally illuminated by, broader issues of substantive historical significance. Given that, I shall reconsider the transcription of each equation to be used in this chapter, beginning in each case with the component equations as Maxwell wrote them, and then proceeding to explicit discussion, as required, of the problems and issues involved in their conversion to vector form.

The preceding discussion of Maxwell's general approach to field theory and his theory of molecular vortices, which together provided the context for the introduction of the displacement current, will give us some purchase for understanding the differences between Maxwell's original treatment of the displacement current and the modern treatment, including the matter of the signs in the equations.[4] It will not be necessary to accuse Maxwell of paying "little . . . attention to signs" and writing "confused equation[s]" in order to explain the differences.[5] Our task will be to see Maxwell's original formulation of the displacement current as he saw it then – as conditioned by nineteenth-century perspectives on field theory – not as a primitive or defective version of our modern formalism.

The standard account of the origin of the displacement current, such as might be encountered in a current textbook or set of lectures on electromagnetic theory, is presented in Section 1 of this chapter. That account will be useful as a comparison template against which the characteristics of the actual historical situation may be delineated by explicit comparison and contrast. The basic claims of the standard account are then tested against the historical record in Sections 2–4. This exercise of comparing the historical record with the standard account will reveal substantial agreement – the standard account is not mere mythology – but also substantial discrepancy – especially in connection with the problem of the algebraic signs. Beginning in Sections 2–4, and more fully in Sections 5–6, the discrepancies will be interpreted by reference to the context in the theory of molecular vortices.

1   *The standard account*

According to the standard account, Maxwell began with a set of four field equations, named, and expressed modernly in macroscopic form, as follows:[6]

Coulomb's law: $$\operatorname{div} \mathbf{D} = 4\pi\rho \qquad (4.1)$$

Ampère's law: $$\operatorname{\mathbf{curl}} \mathbf{H} = 4\pi\mathbf{J} \qquad (4.2)$$

Faraday's law: $$\operatorname{\mathbf{curl}} \mathbf{E} = -\frac{\partial \mathbf{B}}{\partial t} \qquad (4.3)$$

Absence of free magnetic poles: $$\operatorname{div} \mathbf{B} = 0 \qquad (4.4)$$

where $\mathbf{D}$ is the displacement, ρ is the electric charge density, $\mathbf{H}$ is the magnetic field, $\mathbf{J}$ is the electric current density, $\mathbf{E}$ is the electric field, $\mathbf{B}$ is the magnetic induction, $\partial/\partial t$ represents partial differentiation with respect to time, and div and **curl** are differential operators with respect to spatial dimensions. Various units are used in different renditions of the standard account; I have selected electromagnetic units, as being closest, on the whole, to Maxwell's units. The eponymous naming of equations (4.1)–(4.3) is conventional, reflecting the ultimate origin of each equation, rather than its statement in the given form.

The macroscopic field vectors $\mathbf{D}$ and $\mathbf{H}$ in equations (4.1) and (4.2) take into account the effects of electrically and magnetically polarizable media; the charge and current densities ρ and $\mathbf{J}$ used in these equations are therefore the "true" charge and current, not including the "bound" charges and currents attributable to polarizable media.[7] For the paradigmatic case of the charging capacitor circuit, as in Figure 4.1a, the true charge and current are respectively the charge on the capacitor plates, as indicated by the plus and minus signs, and the conduction current in the wire, as indicated by the arrows. (A source of electromotive force to

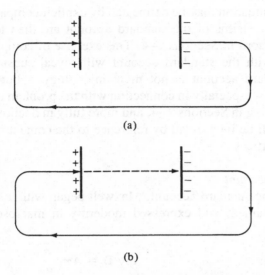

(a)

(b)

Figure 4.1. Charging capacitor circuit, standard account: (a) true charge and current, open circuit; (b) additional current between plates, yielding solenoidal, closed circuit.

produce the current in the circuit is assumed, although not explicitly represented in the diagram.) Apart from the field equations, conservation of charge requires the following:[8]

$$\text{Equation of continuity:} \qquad \text{div } \mathbf{J} + \frac{\partial \rho}{\partial t} = 0 \qquad (4.5)$$

Maxwell realized – the standard account continues – that this was not yet a complete and consistent set of equations: In equation (4.2), Ampère's law, the **curl H** term, by virtue of its mathematical structure, is divergence-free, and it follows that **J** must be divergence-free, which is to say that the equation can be applied only to closed circuits. For open circuits, on the other hand, in which electric charge is accumulating, the divergence of **J** will be nonzero. Thus, consider again the charging capacitor circuit in Figure 4.1a: While the capacitor is charging, both the div **J** and $\partial \rho / \partial t$ terms in equation (4.5), the equation of continuity, will be nonzero. Ampère's law, however, requires that div **J** = 0, and is thus inconsistent with the equation of continuity for the case under consideration – namely, the open circuit. Thus, Ampère's law is in need of modification.

The difficulty could be resolved, Maxwell realized, by adding to the right-hand side of equation (4.2) another term, which would function effectively in the equation as a current density and which, taken together with the true or conduction current density **J**, would yield a composite current that was divergence-free, or "solenoidal."[9] Consider again the charging capacitor circuit, Figure 4.1a. What was required for mathematical consistency in that situation was an additional current between the capacitor plates, as indicated by the broken arrow in Figure 4.1b, that would close the loop, thus giving rise to a divergence-free composite current.

The form of the needed term could be further specified by requiring consistency with equation (4.1), Coulomb's law, which in the simple vacuum case reduces to the following:

$$\text{div } \mathbf{E} = 4\pi c^2 \rho \qquad (4.1')$$

where c is the ratio between electromagnetic and electrostatic units. If a new term proportional to the time derivative of the electric field is introduced into Ampère's law,

$$\mathbf{curl\ H} = 4\pi \mathbf{J} + \frac{1}{c^2} \frac{\partial \mathbf{E}}{\partial t} \qquad (4.2')$$

then, taking the divergence of this equation, we obtain

$$\text{div } \mathbf{J} + \frac{1}{4\pi c^2} \text{div } \frac{\partial \mathbf{E}}{\partial t} = 0$$

which, upon interchanging time differentiation and divergence, and then substituting for div **E** according to Coulomb's law, as in equation (4.1'), becomes precisely the equation of continuity. Thus, the modification of Ampère's law, as in equation (4.2'), achieves a network of consistency among Ampère's law, Coulomb's law, and the equation of continuity.[10]

To return to the example of the charging capacitor circuit, as in Figure 4.2a: With the current in the wire as indicated, the positive charge will be accumulating on the left plate; the electric field **E** will be pointing to the right; **E** being an increasing quantity, its time derivative $\partial \mathbf{E}/\partial t$ will also point toward the right; and the new term, $(1/c^2)\ \partial \mathbf{E}/\partial t$, will be of the appropriate sign and magnitude to close the circuit, thus rendering equations (4.1'), (4.2'), and (4.5) mutually compatible. The conduction current **J** and the new term in $\partial \mathbf{E}/\partial t$ here exist in distinct spatial locations, so that the addition contemplated in equation (4.2') is not an addition of two terms that coexist spatially, but rather indicates the contribution of sources of the magnetic field in spatially distinct areas to the constitution of a closed loop, as diagrammatically represented in Figure 4.2b.[11]

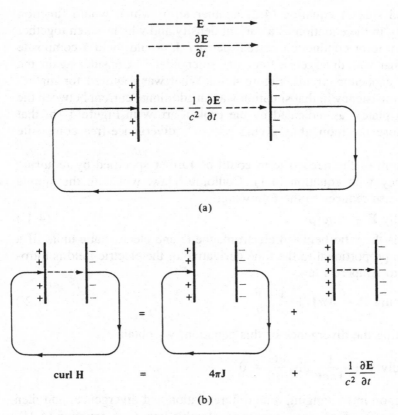

Figure 4.2. Charging capacitor circuit, standard account, $\partial\mathbf{E}/\partial t$ term in Ampère's law: (a) $(1/c^2)\,\partial\mathbf{E}/\partial t$ additional current term between plates; (b) diagrammatic representation of summation in Ampère's law.

Some versions of the standard account end here, presenting the modification of Ampère's law as in essence a mathematical maneuver on Maxwell's part to deal with the problem of the open circuit. Other articulations of the standard account go further, suggesting a role for physical considerations – having to do with the concept of dielectric polarization – in motivating the new term. Going back to the charging capacitor circuit, consider the case in which there is a block of dielectric material in the space between the capacitor plates, as in Figure 4.3a. In this situation, the electric field $\mathbf{E}$ attributable to the accumulating charge on the plates will tend to polarize the molecules of the dielectric as indicated, so that the current attributable to the growing polarization will be in the proper direction to close the loop; if $\mathbf{P}$ is the polarization – or dipole moment per

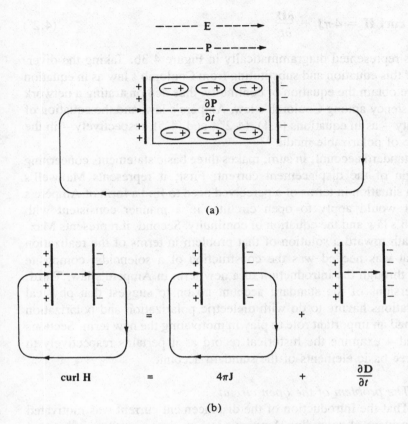

Figure 4.3. Charging capacitor circuit, standard account, dielectric material between plates: (a) $\partial P/\partial t$ additional current term between plates; (b) diagrammatic representation of summation in Ampère's law.

unit volume – the current will be given by $\partial P/\partial t$, with both P and $\partial P/\partial t$ pointing toward the right, as indicated. This physical argument is only suggestive, rather than determinative, as the value of the polarization current will in all cases be less than is required to completely close the loop, and there will be no polarization current in vacuo. We can, however, make use of the macroscopic displacement vector D, which is closely related to the polarization vector P:

$$D = \frac{1}{c^2} E + 4\pi P \qquad (4.6)$$

If we add the displacement current, $\partial D/\partial t$, rather than the polarization current, $\partial P/\partial t$, into equation (4.2), we obtain

$$\mathbf{curl\ H} = 4\pi\mathbf{J} + \frac{\partial\mathbf{D}}{\partial t} \qquad (4.2'')$$

which is represented diagrammatically in Figure 4.3b. Taking the divergence of this equation and substituting from Coulomb's law as in equation (4.1), we obtain the equation of continuity, thus demonstrating a network of consistency among Coulomb's law, Ampère's law, and the equation of continuity – as in equations (4.1), (4.2''), and (4.5), respectively – in the presence of polarizable media.[12]

The standard account, in sum, makes three basic statements concerning the origin of the displacement current: First, it represents Maxwell's problem situation in terms of a perceived need to find a form of Ampère's law that would apply to open circuits, in a manner consistent with Coulomb's law and the equation of continuity. Second, it represents Maxwell's path toward a solution of that problem in terms of the realization that what was needed was the construction of a solenoidal composite current, through the introduction of a new term in Ampère's law. Third, some versions of the standard account go on to suggest that physical considerations having to do with dielectric polarization and polarization current had an important role to play in motivating the new term. Sections 2, 3, and 4 examine the historical record as it pertains respectively to these three basic elements of the standard account.

2   The problem of the open circuit

That the introduction of the displacement current was motivated by the problem of extending Ampère's law to open circuits is supported by the historical record, although, as we shall see, the record does not give this element quite the centrality that it has in the standard account.

In Maxwell's first major paper on electromagnetic theory, "On Faraday's Lines of Force" (1855), he made use of a set of component equations substantially equivalent to the formulation of Ampère's law that is the starting point of the standard account – with some subtle but significant differences. Maxwell wrote the equations as follows:

$$a_2 = \frac{d\beta_1}{dz} - \frac{d\gamma_1}{dy}$$

$$b_2 = \frac{d\gamma_1}{dx} - \frac{d\alpha_1}{dz} \qquad (4.7)$$

$$c_2 = \frac{d\alpha_1}{dy} - \frac{d\beta_1}{dx}$$

where a_2, b_2, c_2 represent "the quantity of electric current at the point (xyz) estimated in the directions of the axes x, y, z respectively," that is, the components of the electric current density **J**, and α_1, β_1, γ_1 are the "effective magnetizing forces," that is, the components of the magnetic field **H**.[13] Thus, for mnemonic and comparative purposes – and without substantial distortion – we may rewrite equations (4.7) in vector form as follows:[14]

$$\mathbf{J} = \mathbf{curl\ H} \qquad (4.8)$$

Comparing this equation with the parallel equation (4.2) of the standard account, we note two differences. First, the constant 4π appears in equation (4.2), but not in this equation; this is just a matter of units, Maxwell having used rationalized units at this point. Second, there is a reversal of left- and right-hand sides, which may appear innocuous, but will turn out to be of some importance.

As correctly suggested by the standard account, Maxwell was fully aware of the limitations of this equation: He was explicit in pointing out that equations (4.7) yield, "by differentiation,"

$$\frac{da_2}{dx} + \frac{db_2}{dy} + \frac{dc_2}{dz} = 0$$

"which is the equation of continuity for closed currents." That is, he recognized that equations (4.7)/(4.8), because they entail zero divergence for the electric current, can apply only to closed current loops: "Our investigations are therefore for the present limited to closed currents."[15] Maxwell apologized for this limitation by alluding to the lack of experimental data concerning the relationship between electric current and magnetic field in open circuits; in that circumstance, the lack of a theoretical treatment was not to be regarded as a pressing issue.[16] Elsewhere in the paper, however, Maxwell had acknowledged that the duties of the theoretician did not end where the experimental data gave out: A proper theory, according to Maxwell, not only "must . . . satisfy those laws, the mathematical form of which is known [from experiment]," but also "must afford the means of calculating the effects in . . . cases where the known formulae are inapplicable."[17] Maxwell's excuse for not fulfilling that duty in "Faraday's Lines" was that he was not proposing a full-blown theory of electromagnetic phenomena in that paper; instead, he was merely preparing the ground for such a theory, through the use of illustrative and heuristic physical analogies. Maxwell did, however, look forward to the time when he would be ready to propose a "mature theory"; presumably, his excuse would expire at that time, and he would then be obligated to confront the problem of the extension of Ampère's law to open circuits.[18]

In "Physical Lines," where Maxwell did present the theoretical rather than analogical treatment to which he had looked forward in "Faraday's Lines" – namely, what he was prepared to identify as the "theory of molecular vortices"[19] – he was presumably to be bound by the standards for such a theory to which he had committed himself – including the requirement that a generalization of Ampère's law, applicable to open circuits, be included in the theory. Maxwell did in fact honor that commitment – in Part III of the paper, if not in Parts I and II. The unmodified form of Ampère's law appeared in "Physical Lines," in component form, as follows:

$$p = \frac{1}{4\pi}\left(\frac{d\gamma}{dy} - \frac{d\beta}{dz}\right)$$

$$q = \frac{1}{4\pi}\left(\frac{d\alpha}{dz} - \frac{d\gamma}{dx}\right) \tag{4.9}$$

$$r = \frac{1}{4\pi}\left(\frac{d\beta}{dx} - \frac{d\alpha}{dy}\right)$$

where "α, β, γ are the rectangular components of magnetic intensity, and p, q, r are the rectangular components of steady electric currents" – that is, respectively the components of the vectors **H** and **J**.[20] The equation can therefore be transcribed (as already done in Chapters 2 and 3 without quite so explicit an analysis) as follows:

$$\mathbf{J} = \frac{1}{4\pi}\ \mathbf{curl\ H} \tag{4.10}$$

This is quite like equation (4.2) of the standard account, with the following differences: The left- and right-hand sides are reversed, just as in "Faraday's Lines"; in addition, the constant 4π that now appears in the equation (unrationalized units) stands not with **J** but, in inverse form, with **curl H**. These do not alter the basic mathematical content of the equation, but they do have definite physical significance, as will be explored in Section 3.

Ampère's law was first modified in Part III of "Physical Lines," where equations (4.9) were recast as follows:

$$p = \frac{1}{4\pi}\left(\frac{d\gamma}{dy} - \frac{d\beta}{dz} - \frac{1}{E^2}\frac{dP}{dt}\right)$$

$$q = \frac{1}{4\pi}\left(\frac{d\alpha}{dy} - \frac{d\gamma}{dx} - \frac{1}{E^2}\frac{dQ}{dt}\right) \qquad (4.9')$$

$$r = \frac{1}{4\pi}\left(\frac{d\beta}{dx} - \frac{d\alpha}{dy} - \frac{1}{E^2}\frac{dR}{dt}\right)$$

where "p, q, r are the electric currents in the directions of x, y, and z; α, β, γ are the components of magnetic intensity; and P, Q, R are the electromotive forces."[21] Let us examine all of the terms of this equation de novo, taking nothing for granted. First, does the vector with components p, q, r – the "electric current" – still represent the electric current density **J**, or does it at this point perhaps represent a composite solenoidal current, such as appears in the standard account?[22] Maxwell's notation certainly suggests that this is still the current density **J**, for he indicated no modification, such as by using a different symbol or adjoining a prime. Furthermore, just after writing this equation, Maxwell went on to substitute it into the equation of continuity. Now, the current that goes into the equation of continuity is precisely the true or conduction current density **J**, rather than any kind of composite solenoidal current, and Maxwell's subsequent manipulations confirm this identification.[23] We must conclude that p, q, r is still to be rendered as the current density **J**.[24] As concerns the vector with components α, β, γ, Maxwell again identified this as the "magnetic intensity," and no ambiguity presents itself in this connection, so that we can safely render this vector as **H**. The **curl** operator is also substantially unproblematic – the substitution of y for z in the second component equation of (4.9') is clearly a slip that is without further significance. Maxwell's P, Q, R, the components of "electromotive force," correspond throughout to the modern electric field **E**, and Maxwell's identification of P, Q, R in the static case with the gradient of the electrostatic potential, a few lines down from equation (4.9'), confirms this identification.[25]

As concerns the constant E, Maxwell's discussion of electrostatic forces (see Chapter 3, Section 4) indicates that this can be mnemonically rendered as the modern ratio of units c. For the vacuum case, Maxwell explicitly identified E as "the number by which the electrodynamic measure of any quantity of electricity must be multiplied to obtain its electrostatic measure." Maxwell's numerical evaluation of E, as "$E = 310,740,000,000 \ldots$ the unit of length being the millimetre, and that of time being one second,"[26] confirms the assimilation of this quantity to the modern ratio of units c, so that Maxwell's equations (4.9') may be rendered, for the case of no dielectric media, and interpreting all of the variables in their electromagnetic rather than their mechanical senses, as follows:

$$\mathbf{J} = \frac{1}{4\pi}\left(\mathbf{curl}\ \mathbf{H} - \frac{1}{c^2}\frac{\partial \mathbf{E}}{\partial t}\right) \tag{4.10'}$$

Comparing now with equation (4.2') of the standard account, we find a basic similarity: By rearranging terms, the two equations can be transformed one into the other; they are mathematically equivalent. Thus far, then, the standard account is vindicated: Maxwell began with an equation mathematically equivalent to equation (4.2) of the standard account (Ampère's law); he was aware of the inadequacies of that equation for the case of the open circuit and therefore wanted to modify it, just as maintained in the standard account; and he accomplished the modification by adding an extra term, arriving at an equation equivalent to equation (4.2') of the standard account, thus achieving consistency with Coulomb's law and the equation of continuity for the open circuit.

On the other hand, if the generalization of Ampère's law to open circuits, for the purpose of achieving a complete and fully consistent set of electromagnetic field equations, had been uppermost in Maxwell's mind, as suggested by the standard account, one would expect that Maxwell would have informed the reader of that, explicitly pointing out the significance in that respect of the new equation. But he did not do that. The discerning reader can read the implicit message carried in the equations: Maxwell's first use of the new form of Ampère's law was to take its divergence, substitute into the equation of continuity, and thereby derive Coulomb's law; that demonstrated the consistency of the new form of Ampère's law with the equation of continuity and Coulomb's law, for all cases, including the open circuit. For Maxwell, in "Physical Lines," however, the main importance of that calculation lay in its relevance to the theory of molecular vortices: The modified Ampère's law, interpreted mechanically, made possible the accumulation of static charge in the theory of molecular vortices, and the calculation utilizing the equation of continuity led to a mechanical interpretation of the electrostatic field [equations (3.13)/(3.14) and schema (3.15)]. It was the theory of molecular vortices that was uppermost in Maxwell's mind, and it was by means of that mechanical representation that Maxwell intended to unify electromagnetic theory; a complete and consistent set of equations would be a necessary part of the exercise, but the center of concern was the mechanical picture, rather than the electromagnetic equations considered in and of themselves, as disembodied mathematical entities. Maxwell accordingly hastened on to develop the implications of the new equation for the theory of molecular vortices, without stopping to comment on its mathematical and phenomenological significance per se.[27]

In sum, Maxwell did basically what the standard account says he did: He modified Ampère's law in order to generalize it to the open circuit, in a manner consistent with the equation of continuity and Coulomb's law. His goal in that, however, was not a complete and consistent set of electromagnetic equations for its own sake, but rather a complete and consistent mechanical model of the electromagnetic field.

3   *The solenoidal current*
The standard account suggests that Maxwell conceptualized the modification of Ampère's law as the construction of a solenoidal current to serve as the source of the magnetic field; the historical record, however, indicates that the idea of the solenoidal composite current appeared only in Maxwell's later work[28] – from which subsequent theory and the standard account stem – and was not a part of the original context of the displacement current. Indeed, a careful reading of the relevant equations as they appear in "Physical Lines" indicates that Maxwell at that time conceptualized the role of the displacement current in a very different way; although the equations are mathematically equivalent to their modern counterparts – as can be seen by rearranging terms and factors – there is, nevertheless, a message implicit in the original arrangement of terms and factors that sheds light on the issue of the solenoidal current.

The difference between Maxwell's approach in "Physical Lines" and the modern approach is signaled at the outset by the manner in which he wrote the unmodified Ampère's law, equations (4.9)/(4.10), in that paper:

$$\mathbf{J} = \frac{1}{4\pi} \text{ curl } \mathbf{H}$$

As has been noted, this differs from the modern form of the equation in two ways: First, the factor 4π is grouped with $\mathbf{J}$ (as $1/4\pi$) rather than with **curl H**; second, the terms are reversed, with $\mathbf{J}$ on the left and **curl H** on the right. The result of these two differences is that $\mathbf{J}$ comes out standing by itself on the left-hand side of the equation. The significance of this can be understood against the background of certain persistent conventions in the writing of equations, one of which has been to put the unknown quantity that is to be calculated – the answer that is to be found – on the left-hand side of the equation, whereas the known or given quantities, which are to give rise to the answer as a result of operations upon them, are written on the right-hand side (at least in the context of the European languages, written from left to right). Considered physically, the right- and left-hand sides in most cases represent cause and effect, respectively: The variables that specify the given conditions or circumstances of the situation stand on the right, whereas on the left stands a variable charac- terizing the outcome of the situation, the effect to which the causes characterized on the right-hand side give rise.[29] Thus, in the modern equation (4.2), the electric current $\mathbf{J}$ is regarded as the "source" of the magnetic field $\mathbf{H}$, and hence appears on the right.[30]

In general, in twentieth-century classical (nonquantum) electrody- namics, charges and currents are regarded as the sources or causes of electric and magnetic fields. One must remember, however, that this way of looking at the relationship between fields and their sources is a rather late addition to the tradition of field theory, stemming from the period around the turn of the twentieth century when H. A. Lorentz combined the Maxwellian tradition, emphasizing the primacy of fields, with the Continental, charge-interaction tradition, which regarded charges and currents as primary; Lorentz thus constructed a dualistic theory in which charges and currents, as well as electric and magnetic fields, are regarded as fundamental, and the former are regarded as the sources or causes of the latter. Lorentz's synthesis became paradigmatic for the twentieth cen- tury.[31]

Maxwell, on the other hand, as a faithful follower of Faraday, was strongly committed to the primacy of the field (see Chapter 1, Section 1). According to Faraday, the hypothesis of electrical particles was objection- able, both because the particles were unobservable and because they constituted a naive reification of electromagnetic phenomena in terms of yet another of the eighteenth-century imponderable fluids. For Faraday, the electric lines of force were primary, and electric charge was simply a manifestation of the terminating points of the lines of force. Associatedly, electric currents were to be regarded as manifesting particular states of the conducting media, conditioned by the surrounding lines of force, rather than transport of electrical fluids or particles. Thus, Maxwell, following Faraday, persistently pursued the goal of casting electromagnetic theory in a form in which the fields would be primary, and charges and currents would be viewed as emergent manifestations of field dynamics. Maxwell was never completely successful at this – charges and currents, as inde- pendent entities, and even as sources of fields, crop up especially in his later work – and the resulting internal conflict is what has made the Maxwellian corpus in electromagnetic theory so difficult to interpret, from Maxwell's day to the present. The prevailing commitment and tendency of Maxwell's field theory were, nevertheless, quite clearly in the direction of regarding the field as primary.[32]

It was in this general context that Maxwell saw fit to characterize the purpose and function of equations (4.9)/(4.10) as "enabl[ing] us to deduce the distribution of the currents of electricity whenever we know the values of α, β, γ, the magnetic intensities."[33] Maxwell thus considered the magnetic field as given, and the electric current as the quantity to be calculated; the **curl H** term therefore appears on the right-hand side of the equation, and the electric current **J** is on the left. The equation was denoted – in places where Maxwell referred to it by name – the "equation . . . of electric currents"; that is, it is an equation that specifies, as its result, the electric current.[34]

In this context, the modification of Ampère's law had to be conceptualized in a manner different from that envisioned in the standard account. Adding the new term to the right-hand side of equations (4.9)/(4.10), Maxwell had arrived at equations (4.9′)/(4.10′):

$$\mathbf{J} = \frac{1}{4\pi} \left(\mathbf{curl\ H} - \frac{1}{c^2} \frac{\partial \mathbf{E}}{\partial t} \right) \qquad (4.10')$$

Here, the summation groups the new term with **curl H** rather than **J**, and the result of the summation is *not* to yield a solenoidal current. The term in $\partial \mathbf{E}/\partial t$ does not serve to close the loop, as in Figures 4.1–4.3; on the contrary, the function of the new term is to open up the circuit, to be added to the solenoidal (closed-loop) **curl H** term (Figure 4.4a) to yield an open circuit for the true current **J**. In other words, the role of the new $\partial \mathbf{E}/\partial t$ term is to cancel out the **curl H** term in the space between the capacitor plates (Figure 4.4b); it can do this because it appears with a negative sign, and hence points toward the positive plate as the capacitor charges up (Figure 4.5a). Interpreting from the field-primacy point of view, in which the fields give rise to currents and charges, rather than vice versa, it would appear that the magnetic field gives rise to a solenoidal current, existing in both the wire and the space between the capacitor plates; that the changing electric field gives rise to a reverse current in the space between the plates; and that the sum of these two partial currents, attributable to two distinct field processes, yields the actual conduction current **J** (Figure 4.5b).

Thus, although Maxwell's modified form of Ampère's law is mathematically equivalent to its counterpart in the standard account, the message of Maxwell's placement and grouping of terms and factors, taken in

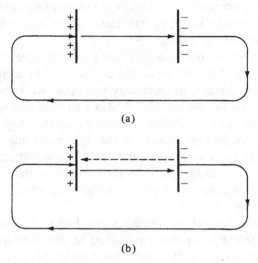

Figure 4.4. Charging capacitor circuit, Maxwell's conception: (a) current owing to solenoidal, closed-loop **curl H** term; (b) additional, reverse current between plates, cancelling **curl H** term, yielding open circuit for true current.

the context of the field-primacy approach to electromagnetic theory, is that he conceptualized the equation in a quite different way. In particular, a solenoidal composite current acting as the source of the magnetic field definitely was not a part of that conceptualization.

4   *The concept of dielectric polarization*

Although the modified form of Ampère's law that appeared in "Physical Lines" involved the time derivative of the electric field – as in the vacuum form of the modern equations – rather than a displacement or polarization variable as in the macroscopic form, Maxwell did invoke the concept of dielectric polarization or "displacement" on his way toward that equation. The concept of displacement was introduced as follows:

Electromotive force acting on a dielectric produces a state of polarization of its parts similar in distribution to the polarity of the particles of iron under the influence of a magnet, and, like the magnetic polarization, capable of being described as a state in which every particle has its poles in opposite conditions. . . .

The effect of this action on the whole dielectric mass is to

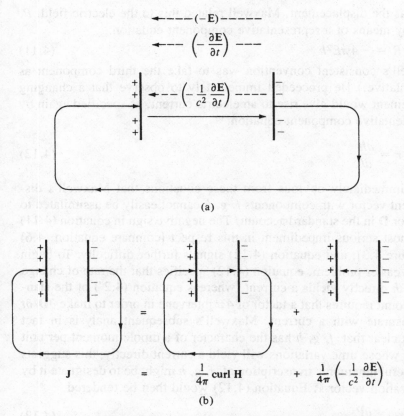

Figure 4.5. Charging capacitor circuit, Maxwell's conception, $\partial E/\partial t$ term in Ampère's law: (a) $[-(1/c^2)\,\partial E/\partial t]$ reverse-current term between plates; (b) diagrammatic representation of summation in Ampère's law.

produce a general displacement of the electricity in a certain direction.

The mathematical paradigm for the treatment of polarized media was Poisson's treatment of the magnetic case; that analysis had been taken over for the electrical case by Ottaviano Mossotti, whom Maxwell cited explicitly; and standing behind Maxwell's physical interpretation of all of this was Faraday's work on dielectrics, also explicitly cited by Maxwell.[35]

The equation that Maxwell wrote for the displacement represented an amalgam of these influences, as well as the requirements of his own specific situation in "Physical Lines." Defining a vector with components

f, g, h as the displacement, Maxwell related this to the electric field, P, Q, R, by means of a representative component equation:

$$R = -4\pi E^2 h \qquad (4.11)$$

(Maxwell's consistent convention was to take the third component as representative.) He proceeded immediately to observe that a changing displacement would give rise to an electric current, as specified again by a representative component equation:[36]

$$r = \frac{dh}{dt} \qquad (4.12)$$

It is immediately obvious from these equations that Maxwell's displacement vector with components f, g, h cannot easily be assimilated to the vector $\mathbf{D}$ in the standard account. The negative sign in equation (4.11) is the most serious impediment in this respect [compare equation (4.6) and Figure 4.3], and equation (4.12) signals further difficulty. To begin with the easier problem, equation (4.12) specifies that the rate of change of f, g, h directly yields a current, whereas equation (4.2″) of the standard account requires that a factor of 4π intervene in order to make $\partial\mathbf{D}/\partial t$ commensurate with a current. Maxwell's subsequent analysis in fact makes it clear that f, g, h has the character of a dipole moment per unit volume, whose time variations will yield a current directly; this suggests that a useful mnemonic transcription of f, g, h might be to designate it by a polarization vector $\mathbf{P}$. Equation (4.12) would then be rendered

$$\mathbf{J} = \frac{\partial\mathbf{P}}{\partial t} \qquad (4.13)$$

In this notation, and making use of our previous mnemonic identifications, equation (4.11) becomes

$$\mathbf{E} = -4\pi c^2 \mathbf{P} \qquad (4.14)$$

Using this mnemonic notation, and making explicit an intermediate step, Maxwell's route to the modification of Ampère's law can be rendered as follows: According to equations (4.12)/(4.13), "a variation of displacement is equivalent to a current," and "this current must be taken into account in [Ampère's law]." When this additional current – the displacement current – is "added to" Ampère's law [equation (4.10)], the latter becomes

$$\mathbf{J} = \frac{1}{4\pi}\,\text{curl } \mathbf{H} + \frac{\partial\mathbf{P}}{\partial t} \qquad (4.10″)$$

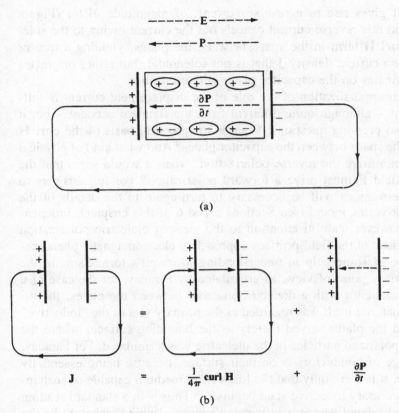

Figure 4.6. Charging capacitor circuit, Maxwell's conception, dielectric material between plates: (a) $\partial P/\partial t$ reverse-current term between plates; (b) diagrammatic representation of summation in Ampère's law.

Immediately substituting from equation (4.14), Maxwell arrived at, and explicitly wrote out, the form of the modified Ampère's law discussed earlier in Section 3 – equations (4.9')/(4.10') – expressed in terms of the time derivative of the electric field, rather than the displacement.[37]

Consider now, once again, the example of the charging capacitor circuit, as in Figure 4.6. Maxwell's conceptualization of this, in terms of the displacement current – and taking into account his emphasis on the primacy of the field – would have been as follows: As the capacitor charges up, there is a growing electric field pointing from positive to negative in the space between the plates; associated with this field there is a *reverse* polarization, as specified by equations (4.11)/(4.14); as this polarization grows, it gives rise to a reverse current, of magnitude $\partial P/\partial t$ (Figure 4.6a), and this reverse current cancels out the current owing to the solenoidal **curl H** term in the space between the plates, yielding a true or conduction current density **J** that is not solenoidal, but rather originates and terminates on the capacitor plates (Figure 4.6b).

This conceptualization of the role of the displacement current is self-consistent – although quite different from the standard account – but it leaves two pressing questions: What is the physical basis of the **curl H** term in the space between the capacitor plates? And what kind of physical process can drive the reverse polarization, when it would seem that the electric field **E** must drive a forward polarization? For full answers to these questions, it will be necessary to turn again to the details of the molecular-vortex model (see Sections 5 and 6 of this chapter). Independently, however, careful attention to the view of dielectric polarization characteristic of the field-primacy approach to electromagnetic phenomena can be of some help in understanding Maxwell's formalism. In the field-primacy point of view, as articulated by Faraday, for the case of a charged capacitor with a dielectric material between the plates, the dielectric material itself was regarded as the primary seat of the "inductive" state, and the plates served merely as the bounding surfaces where the chain of polarized particles in the dielectric was terminated. For Faraday, "all charge of conductors is on their surface, because being essentially inductive, it is there only that the [dielectric] medium capable of sustaining the necessary inductive state begins."[38] Thus – in a standard reading of this and related passages in Faraday's work – what is taken to be the charge on a conductor is really nothing but the apparent surface charge ("bound" charge) of the adjacent dielectric medium: Within the dielectric medium, adjacent polar charges cancel each other, as indicated by the broken lines in Figure 4.7, leaving only the surface layers effective; Poisson's classical analysis for the case of magnetic polarization – which was available to both Faraday and Maxwell – renders this argument quantitative. Interpreting Maxwell's account of the charging capacitor in this way, it is the reverse polarization that gives rise to the charge on the capacitor plates, rather than vice versa. [Indeed, equation (4.10″), and its rendering in Figure 4.6b, would appear to require this interpretation: The **curl H** term is solenoidal, and hence does not contribute to any accumulation of charge; it is only the $\partial P/\partial t$ term that is capable of producing accumulations of charge, and charge accumulated in this manner will necessarily be the dielectric surface charge given by the Poisson analysis.] Viewing the relationship between charge and polarization in this

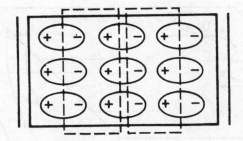

Figure 4.7. Faraday/Maxwell (field-primacy) understanding of reverse polarization in dielectric medium as giving rise to charge on capacitor plates.

way, the justification for the negative signs in equations (4.11)/(4.14) becomes evident.[39]

The notion of the primacy of the field and the attendant notion that electric charge is the manifestation of terminations of lines of dielectric polarization thus go a long way in helping us to understand the particularities of Maxwell's use of the concept of dielectric polarization in motivating the introduction of the displacement current. For a full understanding, however, we must turn again to the molecular-vortex model.

5   *The background in the molecular-vortex model*

The theory of molecular vortices provided the immediate context for the introduction of the displacement current, and for full understanding we must consider again some of the intricacies of the molecular-vortex model. Let us, for this purpose, go back to the penultimate version of the model, as presented in Part II of "Physical Lines," where the original version of Ampère's law,

$$\mathbf{J} = \frac{1}{4\pi} \, \text{curl} \, \mathbf{H}$$

was seen as expressing the relationship between the rotations of the vortices – considered rotating as rigid spheres – and the attendant translational motions of the small particles.[40]

It will be useful at this point to attempt to visualize concretely how this version of the model would function in a particular case; guidelines for the application of the theory to particular cases are provided by the various pictorial representations and associated verbal descriptions that Maxwell included in "Physical Lines."[41] Consider the case of a long straight wire of uniform circular cross section; Figure 4.8 depicts a seg-

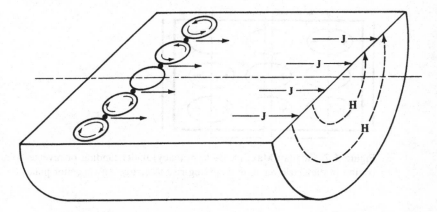

Figure 4.8. Segment of a long, straight conducting wire: on the right, configuration of magnetic field lines **H** and current density **J**; on the left, corresponding patterns of rotation of molecular vortices and translation of idle-wheel particles.

ment of such a wire, sectioned so as to expose the midplane. (As this form of Ampère's law applies only to closed circuits, one must assume that the wire eventually closes back on itself, forming a loop that is large compared with the cross section of the wire, so that the approximation of an infinite straight wire may be used near the wire and inside of it.) Assuming a uniform current density of magnitude *J* in the wire, one can easily solve the differential equation specified by Ampère's law; one finds that within the wire, the magnetic field lines will be circles concentric with the axis of the wire, with the magnitude of the field, *H*, proportional to the distance from the axis, *r*:

$$H = 2\pi J r$$

J and **H** in this configuration are schematically depicted on the right-hand side of Figure 4.8, where the current density **J** is represented by the solid arrows, and the concentric magnetic field lines **H** are represented by the broken arrows. (Outside of the wire, the magnetic lines of force are still circles concentric with the axis, but the field intensity diminishes as the inverse of the distance from the axis.[42])

The molecular-vortex substratum corresponding to this field configuration is depicted on the left-hand side of Figure 4.8 (compare also Figure 4.9a). As the magnetic field lines are perpendicular to the section plane, so also will be the axes of rotation of the vortices, and the motions of the vortices will therefore be rotations of the pseudocircular cross sections in the plane, as depicted by the curved arrows. (The depiction of the vortex cross sections as circular here follows Maxwell's mathematical approximations and the more schematic of his pictorial representations; I think this best captures Maxwell's stance, as he himself was reticent concerning the precise shapes of the vortices – the hexagonal cross sections being merely illustrative.) The vortex on the symmetry axis – where the magnetic field is zero – will not rotate; those off the axis will rotate, with increasing angular velocity as the distance r from the symmetry axis increases, representing the linear variation of the magnetic field intensity; and the rotations will have opposite senses on opposite sides of the symmetry axis, corresponding to the opposite directions of the magnetic field. The effect of the varying rotational velocities will be to push the interposed small particles along in the direction specified by the swifter of the adjacent vortices, as indicated by the straight arrows; the resulting flux density of the small particles then properly represents the electric current density $\mathbf{J}$.[43]

In Part III of "Physical Lines," the substance of the vortex cells was endowed with elastic properties; as we have already seen, the effect of that was, if one may so say, to loosen up the relationship between the vortex rotations and the motions of the small particles – that is, the relationship between magnetic field and electric current: The original relationship – corresponding to the unmodified form of Ampère's law – had been derived in the theory of molecular vortices by assuming that the vortices rotated as rigid spheres; making the vortices deformable altered that relationship, as expressed in the intervention of a new term in the equation [equation (3.12)], where the role of the new term was to "correct the equation . . . of electric currents for the effect due to the elasticity of the medium." That is, elastic deformations of the vortex blobs result in displacements of the surfaces of the vortices; the small particles at the surfaces of the vortices are carried along by these elastic displacements, and these motions of the small particles constitute an additional contribution to the electric current, to be added, in the form of a correction term, to the original **curl H** term – which had been calculated on the assumption of rigid rotation of the vortices, with no elastic distortion. The elasticity of the vortex medium thus provided the mechanical basis for the introduction of the displacement current.[44]

It will be useful, at this point, to go back to our example of the charging capacitor circuit and attempt to visualize concretely how the full model, with elasticity, would function in this situation. Guidelines for application of the theory to this particular case are provided by Maxwell's account of the charged capacitor (or "Leyden jar") and by his general treatment of the accumulation of electric charge, taken in conjunction with his account of conduction currents.[45] We note first that in this ultimate version of the theory, the difference between a conducting medium and an insulating medium was conceptualized as follows: In a conducting medium, the small particles are free to move, and the rotations of the vortices are thus able to set the particles in motion, giving rise in this way to electric currents; in an insulating medium, on the other hand, the particles are not free to move, so that no matter what the motions of the vortices might be, no electric currents can be generated. Consider now a capacitor formed by a thin gap in a thick wire, as in Figure 4.9a, where the wire is shown in section. Surfaces AA and BB define the left and right edges of the gap, respectively, and the gap is filled with a dielectric material. Suppose further that the wire forms a loop – except for this gap – and that there is a source of electromotive force – a battery or a generator – that generates a current in the wire. The connected system of vortices in the wire will have a pattern of rotation just like that in Figure 4.8, and within the wire the small particles will be free to move, and will therefore be pushed along toward the right. The particles in the left-hand lead will accumulate at the edge of the gap – surface AA – where they will be stopped, because they cannot move into or through an insulating medium; in this way a positive electric charge will be built up on surface AA. (The mechanism of the immobilization of the small particles in insulating media has to do with the molecular structure of the insulating material: The material molecules are large structures compared with the molecular vortices, and each material molecule contains many molecular vortices within its boundaries; the small particles are completely immobilized within the material molecules of the insulating medium.) Conversely, the particles in the right-hand lead will be moving away from surface BB, leaving a deficiency, or negative charge, in that place.[46]

Meanwhile, within the thin slab of dielectric material – here represented as comprising only one layer of vortex cells – the vortices are being driven in such a manner as to cause them to undergo elastic distortion. The last row of vortices within the left-hand lead drives the row of vortices within the dielectric, by means of the interposed particles at surface AA, which function as stationary idle wheels (as do all of the small particles within or at the surface of the dielectric); similarly, the first row of vortices within the right-hand lead drives the vortices in the dielectric, by means of the interposed particles at surface BB, which also function as stationary idle wheels. Now, a stationary idle wheel, coupling two wheels of equal size – such as the vortices in cross section – will

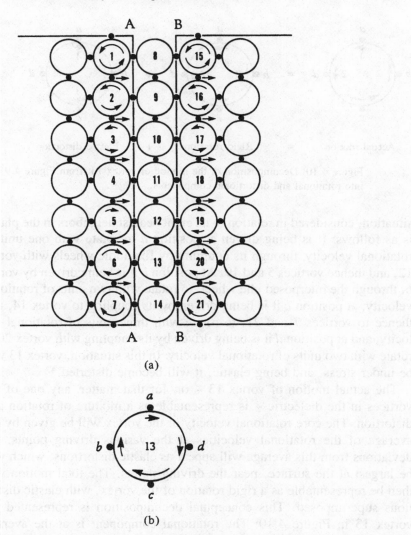

Figure 4.9. (a) Pattern of rotations of molecular vortices at a thin gap, AB, in a thick conducting wire. (b) Detail of vortex 13.

force the two wheels to rotate with equal rotational velocities; any interfering constraints that tend to prevent this equality of rotational velocity will cause stresses on the machinery. In our simplified example, each vortex in the dielectric is being driven through idle wheels at four places, and the driving vortices are moving at various rotational velocities. Consider, for example, vortex number 13 – enlarged in Figure 4.9b – whose

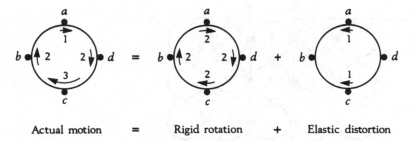

Actual motion   =   Rigid rotation   +   Elastic distortion

Figure 4.10. Decomposition of the motion of vortex 13 (from Figure 4.9) into rotational and distortional components.

situation, considered in relation to its eight nearest neighbors in the plane, is as follows: It is being driven at position *a* to rotate with one unit of rotational velocity, through its coupling by fixed idle wheels with vortex 12, and thence vortices 5 and 19. At position *b* it is being driven by vortex 6, through the interposed idle wheel, to rotate with two units of rotational velocity; at position *c* it is being driven by its coupling to vortex 14, and thence to vortices 7 and 21, to rotate with three units of rotational velocity; and at position *d* it is being driven by its coupling with vortex 20 to rotate with two units of rotational velocity. In this situation, vortex 13 will be under stress, and being elastic, it will become distorted.[47]

The actual motion of vortex 13 – or, for that matter, any one of the vortices in the dielectric – is representable as a mixture of rotation and distortion: The core rotational velocity of the vortex will be given by the average of the rotational velocities at the various driving points; the deviations from this average will appear as elastic distortions, which will be largest at the surface, near the driving points. The total motion will then be representable as a rigid rotation of the vortex, with elastic distortions superimposed. This conceptual decomposition is represented for vortex 13 in Figure 4.10: The rotational component is at the average driven rotational velocity of 2 units – the average of 1, 2, 3, and 2 at *a*, *b*, *c*, and *d*, respectively – and the distortional component then has to be zero at *b* and *d*, and 1 unit of velocity, toward the left, at *a* and *c*.

It is this conceptual breakdown of the motions of the vortices into two components – rotational and distortional – that provides the basis for two kinds of contributions to the electric current: On the one hand, the rotations of the vortices push the small particles along, giving rise to a current of magnitude $(1/4\pi)$**curl H;** on the other hand, the progressive elastic distortions of the vortices move the small particles along in the direction of the distortion, giving another contribution to the electric current – the $\partial \mathbf{P}/\partial t$ term, or displacement current. The significance in the theory of molecular vortices of the modified Ampère's law, equation (4.10″), now becomes clear: The two terms on the right-hand side of the equation designate these two contributions to the electric current: one owing to rotations of the vortices – this is the solenoidal, **curl H** term – and one owing to the elastic distortions of the vortices – this is the displacement current, or $\partial \mathbf{P}/\partial t$ term.

In the case of the charging capacitor circuit, the two terms are in opposite directions in the space between the plates and cancel each other. The capacitor circuit as a whole can thus be conceptualized as follows: As in Figure 4.9a, the vortices within the dielectric are driven, by their coupling with the vortices in the conducting wires, to rotate in a pattern that would – taken by itself – drive the small particles in the forward direction; this gives rise to a **curl H** term in the forward direction within the dielectric, which, taken along with the **curl H** term in the wire, constitutes a solenoidal contribution to the current **J** in the circuit (Figure 4.6b). There is, however, another contribution to the current **J** in the space between the plates. Owing to the immobilization of the small particles in the dielectric, they act as stationary idle wheels, thus constraining the motions of adjacent vortices to conform to each other. These constraints on the motions of the vortices cause elastic distortions of the vortex material that would – taken by themselves – move the small particles in the reverse direction (Figure 4.10); this gives rise to a $\partial \mathbf{P}/\partial t$ term in the reverse direction that cancels out the **curl H** term within the dielectric – compare Figures 4.4b and 4.6b – yielding finally an open current **J**, which begins and ends on the capacitor plates. The two opposing currents in the dielectric medium are in a sense theoretical artifacts: They have their significance in the theory of molecular vortices, but they are destined to cancel each other out and thus give rise to no actual current. In sum, the **curl H** term in the dielectric medium expresses the current that would exist there, owing to the rotations of the vortices, if the small particles were unconstrained; the $\partial \mathbf{P}/\partial t$ term expresses the effect of the constraint, namely, to produce a reverse current, which cancels the **curl H** term. In the final analysis, then, it is the constraint that "drives" the reverse current.

It is thus only by understanding the mechanical basis of equation (4.10″) that we can appreciate the significance of each of its terms for Maxwell and understand why they have the algebraic signs that they do. In particular, we are now able to understand fully why the displacement current had to be, for Maxwell in "Physical Lines," a reverse current, going from the negative toward the positive plate within a charging capacitor, rather than vice versa as in the modern formalism. We are also now able to understand what it is that "drives" this reverse current: It is the constraint on the motion of the small particles, reacting back as a constraint on the motions of the vortices, that drives the elastic distortion of the vortices in the reverse direction. The ramifications of this mechanical conceptualization of the new form of Ampère's law were further developed in Maxwell's treatment of electrostatics.

6   *Displacement and electrostatics*

A primary reason for the introduction of the displacement current was to facilitate the extension of the theory of molecular vortices to embrace electrostatics, and we can gain a fuller understanding of Maxwell's concept of electric displacement by examining its role in his account of electric charge and the electrostatic field. To return to the example of the charging capacitor, the accumulation of charge on the plates is accompanied by elastic distortions of the vortices in the space between the plates. The vortices, in turn, as a result of their elastic deformations, exert reaction forces on their constraints – that is, on the small particles at their surfaces. These kinds of forces, exerted by the vortex surfaces on the small particles, had been discussed earlier in the paper in connection with electromagnetic induction, and Maxwell had identified these forces as "electromotive" forces – forces tending to set the small particles in motion, thus giving rise to electric currents [schema (3.10)]. As we saw earlier (Section 2), these electromotive forces, designated by Maxwell in terms of components $P, Q, R,$ are equivalent, in their general electromagnetic significance, to the modern electric field $\mathbf{E}$. In the case of the charged capacitor (Figure 4.11), the elastic distortions, and hence the polarization $\mathbf{P}$, point toward the positive plate; the elastic restoring forces or electromotive forces $\mathbf{E}$ point in the opposite direction, away from the positive plate. $\mathbf{E}$ thus points in the direction in which the small particles would move, under the influence of the elastic reaction forces, in case of dielectric breakdown. This furnishes the mechanical basis, in the static case, for the negative sign in the relationship between $\mathbf{E}$ and $\mathbf{P}$, as in equation (4.14).[48]

More generally, in Maxwell's theory, any accumulation of charge is accompanied by a pattern of elastic deformation in the magnetoelectric medium, for it is only by the elastic deformation of the vortices that the

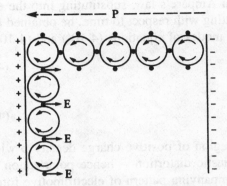

Figure 4.11. Polarization **P** and electromotive force **E** between the plates of a charged capacitor in the molecular-vortex theory.

solenoidal property of the electric current is relaxed, allowing for the accumulation of electric charge through progressive distortion of the medium. The resulting pattern of elastic distortion is in turn accompanied by a pattern of elastic restoring forces, corresponding to the electric field **E**. This is the basis, in the theory of molecular vortices, for the connection between electric charge and the electric field: An electric charge is a center of elastic deformation in the surrounding space, and this elastic deformation in turn gives rise to a pattern of electromotive forces, which is the electric field [see also schema (3.15)].

Maxwell's first use of his new form of Ampère's law was to develop this connection between electric charge and electric field. Beginning with the new equation [mnemonically transcribed herein as either equation (4.10′) or equation (4.10″)],

$$\mathbf{J} = \frac{1}{4\pi} \, \text{curl } \mathbf{H} + \frac{\partial \mathbf{P}}{\partial t} \qquad (4.10'')$$

or, in terms of **E**,

$$\mathbf{J} = \frac{1}{4\pi} \left(\text{curl } \mathbf{H} - \frac{1}{c^2} \frac{\partial \mathbf{E}}{\partial t} \right)$$

Maxwell next wrote the equation of continuity (the transcription being unproblematic),

$$\text{div } \mathbf{J} + \frac{\partial \rho}{\partial t} = 0$$

whence evaluating div **J** from Ampère's law, substituting into the equation of continuity, and integrating with respect to time, he obtained [continuing the dual transcription modes of equations (4.10′) and (4.10″)]

$$\rho = -\text{div } \mathbf{P} \tag{4.15}$$

or

$$\rho = \frac{1}{4\pi c^2} \text{ div } \mathbf{E} \tag{4.16}$$

where the result is that any region of positive charge density ρ will be surrounded by a pattern of elastic distortion – hence polarization **P** – pointing inward, and an accompanying pattern of electromotive force **E** pointing outward (Figure 4.12). These last equations appear in characteristic form – with the field quantities on the right – confirming the perspective that the elastic distortion, which is the mechanical basis of the electric field, gives rise to the electric charge, rather than vice versa.[49]

In all of this, the algebraic signs, and the physical conditions they represent, are completely consistent and coherent within the context of Maxwell's theory, although quite different from the standard account. There was, however, one part of Maxwell's presentation in which there was a certain amount of ambiguity and confusion in connection with the algebraic signs. The source of that was not any conceptual confusion on Maxwell's part, but rather a didactic strategy gone astray. In introducing the concept of dielectric polarization, and in investigating the patterns of elastic stress and strain in the vortex medium that would be associated with dielectric polarization, Maxwell first considered the elastic distortions of the vortices, and the attendant polarizations, in isolation from the rotations of the vortices – no doubt with the intention of simplifying the discussion.[50] The elastic distortions and the rotations were to be put back together later on.[51] The problem with this didactic strategy was that so long as the polarization process was being considered in isolation from the rotations of the vortices, there was no mechanism in view to drive the reverse polarization. This led to a certain awkwardness in Maxwell's presentation.

The awkwardness began with the introduction of the concept of dielectric polarization, where the statement that "electromotive force acting on a dielectric produces a state of polarization of its parts" had been juxtaposed with equations (4.11)/(4.14),

$$\mathbf{E} = -4\pi c^2 \mathbf{P} \tag{4.14}$$

where **E** represents electromotive force and **P** represents polarization.

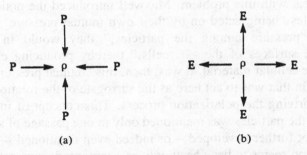

Figure 4.12. Field patterns surrounding a region of positive charge density ρ: (a) polarization **P**; (b) electromotive force **E**.

Maxwell thus seemed to be saying, in a very clear and direct way, that electromotive force acting on a dielectric would produce reverse polarization, and that simply does not make sense. I think we can see Maxwell's dilemma: On the one hand, he knew that he would need the negative sign later on, for consistency in the broader context of the full theory of molecular vortices, where the rotations produce the reverse polarization **P**, and the reaction force **E** is in the forward direction. On the other hand, he wanted at this point to introduce the concept of dielectric polarization in terms of a simple physical picture, independent of the full particulars of the theory of molecular vortices. The result was a mismatch between equation and physical picture.[52]

Even within the context of the simplified picture of dielectric polarization, Maxwell did make some attempt to justify the negative sign in equation (4.14), by discussing reaction forces:

> According to our hypothesis, the magnetic medium is divided into cells, separated by partitions formed of a stratum of particles which play the part of electricity. When the electric particles are urged in any direction, they will, by their tangential action on the substance of the cells, distort each cell, and call into play an equal and opposite force arising from the elasticity of the cells.

This discussion does justify the negative sign in equation (4.14), by interpreting the electromotive force **E** in that equation – in accordance with Maxwell's consistent usage – as the "equal and opposite" reaction force; still, in the absence of the rotations of the vortices, it is not clear by what force the electrical particles are "urged in [some] direction" in the first place, so as to call forth the reaction forces.[53]

In order to deal with this problem, Maxwell introduced the notion of the small particles "being acted on by their own mutual pressure"; as a result of this "pressure among the particles," they would in turn "act . . . on the surfaces of the . . . cells," thereby producing elastic distortions of the cellular material. It was, then, this "mutual pressure" of unspecified origin that was to act here as the surrogate of the rotations of the vortices in driving the polarization process. This concept of mutual pressure among the particles was mentioned only in one passage of seven lines and was not further developed – or indeed even mentioned – elsewhere; its function seems to have been only as a temporary surrogate for the action of the rotations, and it did not find a place in the full theory.[54]

This surrogate electromotive force was used in Maxwell's Proposition XIII, where the much-noted error in sign is found. At the beginning of the proposition, the quantity E (with components P, Q, R) was taken to represent the surrogate electromotive force; regarded as the cause of the polarization P (with components f, g, h), E necessarily pointed in the same direction as P.[55] By the end of the proposition, however, as the result of an error in sign, Maxwell had regained equation (4.11), with a negative sign in the relationship between E and P.[56] That error in sign was, in turn, compensated for by having the quantity E lose its temporary meaning as the surrogate polarizing force, reverting to its otherwise consistent meaning, throughout "Physical Lines," as the force exerted by the vortices on the small particles. That amounted to two changes of sign, which canceled each other out, leaving Maxwell with the result that he needed for coherence in the larger context. The rotations and elastic distortions of the vortices were thereafter put back together in the modified Ampère's law, and all need for (or mention of) the surrogate polarizing force thereby disappeared. Thus, the basic coherence of the theory as a whole was not compromised by the episode of the surrogate polarizing force; the damage remained confined to the perennially perplexing Proposition XIII.

The result of Maxwell's analysis of the mechanical aspects of the polarization process was that the originally spherical vortices would deform – in response to either the surrogate polarizing force or the action of the system of rotating vortices – as in Figure 4.13a, where the broken lines represent radii of the undistorted spherical vortex, and the solid lines represent the distortion of these; the symmetry axis is along the direction of the E and P vectors. According to Maxwell's assumptions and calculations, then, the surface of the vortex would remain spherical, and points on the surface would be displaced along a meridian, with the magnitude of the displacement being proportional to the sine of the polar angle, and

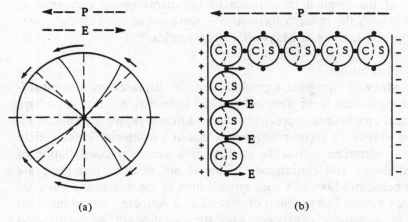

Figure 4.13. Patterns of elastic deformation: (a) a single vortex; (b) an array of vortices between the plates of a charged capacitor.

hence greatest at the equator, as indicated by the arrows. The result would be a net displacement of the particles at the surface in the reverse direction.[57]

Returning again to the example of the charging capacitor, the vortices between the plates become deformed – as a result of their rotation against constraint – as depicted in Figure 4.13b, where the deformation of the equatorial plane of each vortex has been indicated by a broken line, and the symbols C and S indicate respectively the portions of each vortex that have been compressed or stretched. It is this elastic deformation of the vortices that is the mechanical basis of the electric field in Maxwell's molecular-vortex model, and the energy of deformation is the electric field energy; Maxwell used this to derive the electrostatic force law in his model. In that way, Maxwell achieved a complete account of electrostatics, including electrostatic forces, fields, energies, and charges, as well as dielectric media.[58]

Maxwell's account of electrostatics, taken as a whole, thus reflected a coherent realization, in mechanical terms, of the field-primacy approach to electricity and magnetism. Moreover, Maxwell's field-primacy electrostatics both justified and required the negative sign in the relationship between electric field and dielectric polarization. Maxwell's primary goal in introducing the displacement current into Ampère's law had been to extend the theory of molecular vortices to electrostatics, and the peculiarities of the original formulation of the displacement current – as compared with the modern formalism – were bound up with the corresponding peculiarities of Maxwell's electrostatics.[59]

Conclusion

Maxwell's original formulation of the displacement current differed in significant ways from our modern formulation – that is perhaps our central conclusion; appreciating these differences, we are able to understand Maxwell's accomplishment as that of a nineteenth-century field theorist, undertaken within the context of a particular constellation of physical theory and commitment. This is not to deny that there are bridges between Maxwell's conceptualization of the matter and our contemporary views: The problem of generalizing Ampère's law to the open circuit in a manner consistent with the equation of continuity and Coulomb's law was a central concern for Maxwell, just as it is for writers of modern textbooks on electricity and magnetism; the idea of dielectric polarization and polarization current provided a suggestive image of electric current in insulating media that was important for Maxwell, just as it is for modern lecturers on electromagnetic theory. These persistent elements in the conceptualization and justification of the displacement current were, however, interpreted in a very different way in Maxwell's original formulation. In the first place, both the problem of modifying Ampère's law and the model of dielectric polarization were considered by Maxwell within the context of his commitment to the primacy of the field. From that vantage point, Maxwell viewed the modification of Ampère's law, not as the construction of a solenoidal current to act as the source of the magnetic field, but rather as the construction of a nonsolenoidal combination of field processes to function as the basis for the true electric current. Also, from the field-primacy vantage point, Maxwell viewed dielectric polarization and displacement current not as effects of electric charge, vectorially oriented away from positive charges, but rather as causes of electric charge, vectorially oriented toward concentrations of positive charge. Furthermore – and again in contradistinction to modern theory – Maxwell conceptualized electromagnetic phenomena in terms of a mechanical substratum: The displacement current made its initial appearance as a term correcting for the elasticity of the vortex medium, and the resulting summation of terms in Ampère's law was interpreted as reflecting contributions to the electric current stemming from both rotations and elastic deformations of the vortices; the addition of terms in the modified Ampère's law thus represented the summation of components of the true current to yield the true current, rather than the summation of true current and displacement current to yield a composite solenoidal current.

By understanding the peculiarities of Maxwell's original formulation of the displacement current as marks of the nineteenth-century context, rather than as defects by comparison with our modern formulation, we are able to appreciate the extent to which Maxwell's initial treatment of the displacement current was internally consistent, although quite different from a modern treatment. Nevertheless, as Maxwell himself was soon to decide, there were certain problems and awkwardnesses in his initial formulation of the displacement current; he was to remedy these in subsequent reformulations, which will be discussed in Chapter 6.

5

The origin of the electromagnetic theory of light

The molecular-vortex model provided the context for the first appearance of both the displacement current and the electromagnetic theory of light. The first form of the electromagnetic theory of light – which differs from the modern form no less than the first form of the displacement current differs from its modern counterpart – appeared in Part III of "Physical Lines," published in January 1862. In brief, the newly introduced elastic property of the magnetoelectric medium allowed for the propagation of transverse shear waves in that medium; calculating, from the parameters of the model, the velocity of such waves – and finding close agreement with the measured velocity of light – Maxwell identified these waves in the magnetoelectric medium as light waves, and he concluded that the magnetoelectric and luminiferous media were one and the same. The broad nineteenth-century background bearing on such a connection between electromagnetism and light is taken up in Section 1 of this chapter. Section 2 then broaches the question of the precise role of the molecular-vortex model in the origin of the electromagnetic theory of light: Was the electromagnetic theory of light, as textbook accounts might suggest, from the outset basically a matter of deriving wavelike solutions from the equations of electricity and magnetism – in which case the mechanical model could have played at most an ancillary role in the genesis of the theory – or did the molecular-vortex model in fact play a more essential role in the initial formulation of the electromagnetic theory of light, as suggested by the fact that the theory made its first appearance in Part III of "Physical Lines"? Section 3 develops an answer to this question – namely, that the molecular-vortex model played a determinative role in the genesis of the electromagnetic theory of light, which conclusion follows from an analysis of the role of the ratio of electrical units in determining the parameters of the molecular-vortex model, and hence the velocity of wave propagation in the magnetoelectric medium. Section 4 presents an analysis of the approximations and adjustments that Maxwell made in setting the parameters of the molecular-vortex model. Section 5 delineates the research program that Maxwell established on the basis of his identification of the magnetoelectric and luminiferous media.

1 *Electromagnetism and light – the nineteenth-century background*

That there was some connection between electrical or magnetic phenomena and light was not a new idea with Maxwell. The idea had been broached in a variety of contexts in the seventeenth and eighteenth centuries and received new impetus in the early decades of the nineteenth century: With the establishment of the wave theory of light, the existence of a luminiferous ether became widely accepted, and speculation was immediate that this ether might have functions other than the propagation of light – including, perhaps, electrical or magnetic functions. In addition, the discovery of the magnetic effects of electric currents by Oersted both reflected and validated the growing belief in the scientific community that all physical phenomena were connected – with the particular implication that electromagnetic phenomena and light might be connected. Thus, speculations concerning the functions of the luminiferous ether and ideas of connectedness centering on Oersted's discovery converged in the early 1820s to highlight the possibility of a connection between optical and electromagnetic phenomena. A concrete model of such a connection was suggested by Ampère in the early 1820s – he suggested that a combination of positive and negative electricity constituted an ether that functioned to transmit electrical and magnetic forces as well as light and heat – and the idea of an etherial connection for electricity and magnetism gained wide currency.[1]

Michael Faraday embraced the idea of the connectedness of all with all, and he undertook experiments, beginning in the 1820s, to find a connection between electricity and light. Faraday, however, objected to hypothetical ethers, and when in 1845 he established a clear experimental connection between electromagnetic phenomena and light – namely, the Faraday rotation – he conceptualized that connection in a manner that did not invoke an ether: For Faraday, the lines of electric and magnetic force were to be regarded as the primary entities, and he suggested, in 1846, that light perhaps consisted of vibrations in these lines of force. William Thomson's private writings make it clear that, for him, the connection between electromagnetic phenomena and light, as demonstrated by the Faraday rotation, was to be understood in terms of a universal ether underlying electromagnetic, optical, and indeed all physical phenomena; Thomson's published writings, however, were somewhat more circumspect on this issue, emphasizing the implications of the Faraday rotation for the rotatory nature of magnetism, rather than for any particular mode of unification of electromagnetism and optics. For Maxwell, then, the clear message from Faraday and Thomson was that there was a definite connection between electromagnetism and light and that it implied a rotatory nature for magnetism; the question of the relevance of all of this to ether theory, however, was unresolved in the legacy of Faraday and Thomson.[2]

The Continental, charge-interaction tradition in electricity and magnetism, as it developed in the 1840s and 1850s, also provided material suggestive of a connection between electromagnetic and optical phenomena. The force law for the interaction of electric charges enunciated by Wilhelm Weber in 1846 accounted for both electrostatic and electromagnetic forces, and so it perforce contained within it a constant, c_W, relating the two kinds of units, as in equation (1.2):

$$F = \frac{ee'}{r^2}\left(1 - \frac{v^2}{c_W^2} + \frac{2ra}{c_W^2}\right) \tag{1.2}$$

Given Weber's definition of units (he used electrodynamic rather than electromagnetic units), his c_W differs from the constant c of modern electromagnetic theory (Maxwell's E or v) by a factor of the square root of 2:

$$c_W = \sqrt{2}c$$

In order to render his theory complete, Weber needed to measure the constant c_W, which involved measuring the electrostatic and electromagnetic forces attributable to a given electric charge, respectively at rest or in motion. That, however, was not a straightforward task, and it was not until 1856 that Weber and Rudolph Kohlrausch were able to announce a definitive result for the ratio of units c_W:

$$c_W = 439{,}450 \times 10^6 \tag{5.1}$$

where the units of the ratio itself are those of a velocity – millimeters per second – because the electromagnetic (or electrodynamic) unit refers to charge interacting by virtue of its velocity, whereas the electrostatic interaction involves no velocity. For Weber, the primary physical significance of the constant c_W was that it specified a relative velocity between two electric charges, such that, at that velocity, their reciprocal electrostatic forces exerted on each other – as specified by the first term on the right-hand side of equation (1.2) – would be exactly balanced by the reciprocal electromagnetic forces – as specified by the second term – causing the net force on each to vanish. Weber did observe that the measured value of c_W was of the same order of magnitude as the velocity of light, but given the numerical discrepancy of a factor of about 3/2, as well as the perhaps more important physical discrepancy – c_W was interpreted as the velocity of a moving charge, whereas the velocity of light was that of a propagating wave – Weber did not see much of significance in the very approximate numerical coincidence.[3]

The question of the bearing of the ratio of units on a possible connection between electromagnetism and light was sharpened in two publications by Gustav Kirchhoff in 1857, in which Weber's force law was used to calculate the propagation of wavelike variations of electric current in conductors. Kirchhoff concluded that in the limit of thin wires and high conductivity, such wavelike disturbances would propagate with velocity

$$V = \frac{c_W}{\sqrt{2}}$$

where for c_W he used the measurement of Weber and Kohlrausch, as in equation (5.1), which he expressed as rounded to 4.39×10^{11}, in millimeters per second; his result was a speed of propagation of 41,950 German miles per second for these waves (which converts to 3.1074×10^8 meters per second). That is, as Kirchhoff explicitly noted, these wavelike disturbances on thin wires would propagate with a "velocity very closely approximating [lit. very nearly equal to – *sehr nahe gleich*] the velocity of light in empty space."[4] Kirchhoff, however, like Weber, did not see the numerical coincidence as indicative of a physical connection. Kirchhoff's "extraordinary failure" to make anything of what was, in his case at least (as opposed to Weber's), a very close numerical coincidence has received much attention, and a variety of explanatory factors have been discussed: (1) Kirchhoff, for the same reasons as Weber, saw c_W (and parameters derived from it) primarily as characterizing a limiting velocity at which the force between two charges would disappear (thus something physically unlike a wave velocity); (2) Kirchhoff's calculation for waves of electric current yielded a velocity corresponding closely to that of light in vacuum only for the high-conductivity case, which was physically most unlike the vacuum situation; (3) Kirchhoff was in general in the habit of viewing the various mathematical analogies found to exist between disparate physical phenomena as formal analogies, carrying no physical or ontological messages; (4) Kirchhoff's methodology in general leaned strongly toward the phenomenological, abjuring the kinds of bold hypotheses that might connect apparently distinct physical phenomena; (5) Kirchhoff saw himself as a German civil servant, whose mission was to carry out his work thoroughly and soberly, rather than to play with ideas. Whatever the specific reasons for Kirchhoff's "extraordinary failure," his case serves to demonstrate that the numerical relationship between the ratio of units and the velocity of light, though certainly suggestive, nevertheless did not, in and of itself, generate any universally compelling imperative toward the unification of electromagnetism and optics.[5]

Thus, neither from the work of Weber and Kirchhoff relating to the ratio of units nor from the ideas of Faraday and Thomson on a connection between electromagnetism and optics would Maxwell have carried away any definite imperative toward the unification of electromagnetic theory and optical theory on the basis of a single ether. Those ideas and results, along with the general nineteenth-century background that they reflected and articulated, were all available to Maxwell (although he may have become aware of Kirchhoff's work only later), and they certainly would have inclined Maxwell to be sensitive to the possibility of such a unification. The proximate motivation and specific program for the unification of electromagnetism and optics were to come, however, from imperatives generated by Maxwell's own developing research program, as centered in the early 1860s on the molecular-vortex model.[6]

2   *The role of the molecular-vortex model*

In attempting to characterize precisely the role of the molecular-vortex model in motivating and facilitating Maxwell's unification of electromagnetism and optics, various possibilities suggest themselves: First, the model may have been merely illustrative. According to this point of view, the unification of electromagnetism and optics came through a direct reduction of optics to electromagnetism; the mechanical model functioned merely to motivate this result, or to illustrate it for the nineteenth-century audience. Second, the model may have been accommodative. In this view, Maxwell showed that a model that originally applied only to electromagnetism could be adjusted – as by the manipulation of parameters – so as to accommodate optical phenomena as well. Third, the model may have been determinative, as will be argued herein. In this view, Maxwell showed that a model adequate to account for electromagnetic phenomena would, without further adjustment, account for optical phenomena as well.

The three positions sketched are in the nature of ideal types and may be taken as representative points along a continuum: Viewing the model as merely illustrative represents one end of the continuum, where the model is seen as playing little substantive role in the development of the scientific result, having only a rhetorical function; viewing the model as determinative defines the other end of the spectrum, where the model is seen as determining or demonstrating the scientific result; and viewing the model as accommodative defines a point in the middle of the spectrum, where the model is seen as rendering significant help along the way to the result, but not determining it. As we are dealing here with a central, indeed paradigmatic, example of nineteenth-century innovation in the context of a mechanical model, the elucidation of this issue will have direct bearing on our general views concerning the role of mechanical models in nineteenth-century science: Did those models merely provide the background for scientific innovation, or did the models in fact play central roles in the innovative process? To what extent were the central features of those nineteenth-century mechanical models merely ad hoc to preexisting experimental results and theoretical commitments, and to what extent did the models themselves stimulate or bring forth new theoretical insights and new directions for experimental research?[7]

In a certain class of accounts of the origin of the electromagnetic theory of light, the molecular-vortex model is seen as having played a relatively minor role – primarily as illustrative or motivational – and Maxwell's procedure is described generally as follows: Having introduced the displacement current, thereby arriving at a complete and consistent set of field equations, Maxwell noticed that these equations could be combined to yield a wave equation for either the electric field E or the magnetic field H; he was then able to obtain solutions for these equations in the form of transverse electromagnetic waves – alternating E and H fields – propagating with a velocity given by the ratio between electromagnetic and electrostatic units. As the velocity specified in that way by the electromagnetic equations turned out to be quite close to the measured velocity of light, Maxwell was able to conclude that light must consist of these transverse electromagnetic waves. In this view, Maxwell's identification of light as electromagnetic waves proceeded directly from the electromagnetic equations, and any mechanical modeling of those equations would have been superfluous at that point; it might have comforted Maxwell or his readers, but it could have added nothing essential to the argument, for the heart of the argument lay in the electromagnetic equations. In this view, then, the molecular-vortex model was excess baggage, quaint nineteenth-century decoration accompanying the equations that were the heart of the matter; the model may have played a role in the introduction of the displacement current or the calculation of parameters, but the derivation of propagating waves proceeded directly from the electromagnetic equations.[8]

Even a cursory look at the historical documentation, however, shows that such an account cannot be maintained and that the mechanical model must have played a more substantial role. In fact, Maxwell's original formulation of the displacement current in "Physical Lines" did not allow for the propagation of electromagnetic waves; thus, the theory of light presented in "Physical Lines" was not a true electromagnetic theory of light – in the sense of identifying light waves as electromagnetic waves – but rather what has been called an "electro-mechanical" theory of light.[9] In a true electromagnetic theory of light, as presented in a modern textbook, the propagation of electromagnetic waves in free space is derived primarily from two of Maxwell's equations: One of these is the modified Ampère's law, which may be expressed for free space in a convenient form by setting the conduction current **J** equal to zero in equation (4.2′) of the standard account:

$$\operatorname{curl} \mathbf{H} = \frac{1}{c^2} \frac{\partial \mathbf{E}}{\partial t} \tag{5.2}$$

The other equation is Faraday's law of electromagnetic induction, as expressed in equation (4.3) of the standard account, with the magnetic field **H** substituted for the magnetic induction **B** (given that $\mu = 1$ in free space):

$$\operatorname{curl} \mathbf{E} = -\frac{\partial \mathbf{H}}{\partial t} \tag{4.3}$$

In qualitative terms, equation (5.2) says that a changing **E** field will give rise to an **H** field in its neighborhood; that **H** field, as it varies, will, according to equation (4.3), give rise to another **E** field, and so on, in a series of leapfrogging **E** and **H** fields constituting a propagating electromagnetic wave. Mathematically, one may combine equations (5.2) and (4.3) to obtain a wave equation in either **E** or **H**, with solutions in the form of waves propagating in vacuum at velocity c.[10]

Maxwell, in "Physical Lines," had close mathematical equivalents of equations (5.2) and (4.3), but he never wrote down anything quite like equation (5.2), and he did not, anywhere in "Physical Lines," combine two such equations to get a wave equation. In the first place, the arrange-

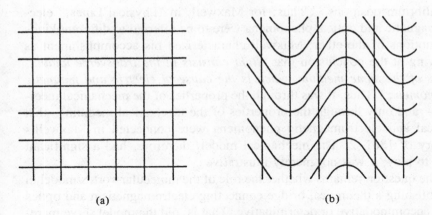

(a)   (b)

Figure 5.1. Behavior of the elastic ether: (a) undisturbed ether; (b) transverse torsion wave in ether. [Note: The vertical lines are parallel in (b) as in (a); look obliquely from the bottom of the page to see this.]

ment of terms in equation (5.2) suggests that the displacement current can function as a source of the magnetic field, and that would have been contrary to Maxwell's commitment to the primacy of the field, in which fields give rise to charges and currents, rather than vice versa. He therefore did not arrange terms as in equation (5.2). Relatedly, because he did not perceive a changing electric field, or displacement current, as being the source of, or giving rise to, a magnetic field, he would not have subscribed to the physical argument of leapfrogging $\mathbf{E}$ and $\mathbf{H}$ fields, or to the corresponding mathematical manipulations leading to a wave equation.

Furthermore, Maxwell conceptualized both electromagnetic processes and optical phenomena in mechanical terms, in a manner that precluded identifying light as a pattern of alternating electric and magnetic fields. Thus, in bringing light into the theory, he viewed light in traditional nineteenth-century mechanical terms – as a transverse torsion wave in an elastic ether (Figure 5.1) – and attempted to relate that picture of light to his electromagnetic theory.[11] The displacements and motions of the ether that take place in the propagation of transverse torsion waves are, however, different from the motions and displacements that correspond to electric and magnetic fields: Magnetic and electric fields correspond to rotations and distortions of molecular portions of ether *in place*, whereas transverse torsion waves correspond to oscillations of larger regions of the ether, involving the motions of whole vortex cells away from their equilibrium positions.[12] Thus, for Maxwell, in "Physical Lines," electromagnetic and optical phenomena were to be referred to different kinds of motions of the ether; Maxwell characterized his accomplishment as arriving at the conclusion that *"light consists in the transverse undulations of the same medium which is the cause of electric and magnetic phenomena."* Thus, it was through the properties of the mechanical medium – and only through the properties of the mechanical medium – that optical and electromagnetic phenomena were connected in Maxwell's theory of 1861–2. The mechanical model, therefore, had a significant role to play; it was not merely illustrative.

The question remains whether the role of the molecular-vortex model in establishing a theoretical bridge connecting electromagnetism and optics was accommodative or determinative. That is, did the model serve merely to show that one could *impose* both electromagnetism and optics on one mechanical model, thus *accommodating* both within one mechanical theory? Or did the model perhaps accomplish something more, namely, to demonstrate that a mechanism adequate to the representation of electromagnetic phenomena would *naturally* or even *necessarily* entail optical phenomena? At first look, the texts seem to provide material in favor of the former position, suggesting that Maxwell imposed the coexistence of electromagnetism and optics on the model by setting the mechanical parameters of the magnetoelectric medium ad hoc to the accommodation of light in the theory.[13]

To begin with, carrying forward a result from Part I of "Physical Lines," Maxwell set the mass density of the medium, ρ_m, to a value of order 1:

$$\mu = \pi \rho_m \tag{5.3}$$

where μ, the magnetic permeability, would be 1 for air or vacuum, so that, for this case, ρ_m would have numerical value

$$\rho_m = \frac{1}{\pi} \tag{5.4}$$

Then, based on calculations in Part III of "Physical Lines," the torsion modulus m was set to a value of order c^2:

$$c^2 = \pi m \tag{5.5}$$

so that m would have the value

$$m = \frac{c^2}{\pi} \tag{5.6}$$

Then, by mechanical considerations alone, the velocity of propagation of transverse torsion waves in the medium, V, would be given by

$$V = \sqrt{\frac{m}{\rho_m}} \qquad (5.7)$$

which, on substituting from equations (5.3) and (5.5), yields exactly

$$V = c$$

That is, the velocity of elastic waves in the magnetoelectric medium is given precisely by the ratio of units, and the close agreement of the latter with the measured velocity of light furnishes the basis for identifying these waves in the magnetoelectric medium as light waves.[14]

It must be admitted that this reasoning has a contrived look: The density of the medium has been set to a number of order 1, the torsion modulus has been set to a number of order c^2, and the factors of π cancel out, so that the square root of the ratio comes out to be precisely c. One certainly comes away with the feeling that the model was adjusted so as to yield that result, thus playing an accommodative rather than a determinative role in the unification of electromagnetism and optics. As I shall proceed to argue, however, the setting of the basic mechanical parameters of the medium was not ad hoc to accommodating light – as appears to be the case on superficial analysis – but grew naturally and independently out of electromagnetic considerations, as a more detailed study of the matter will demonstrate.

3   *The ratio of units and the parameters of the medium*

Our concern now shifts to the following question: What was the rationale for equations (5.3) and (5.5), which established numerical values for the medium constants ρ_m and m, respectively? In particular, what was the reason for introducing the ratio of units c into equation (5.5) for the torsion modulus m of the medium? Was that an arbitrary and imposed manipulation of units, accomplished by switching from electromagnetic units to electrostatic units at just the right moment, or did it grow naturally out of the circumstances of the model? In order to make a judgment on this question, it will be necessary to undertake a critical analysis of the setting of all of the basic parameters of the model, including the mass density of the magnetoelectric medium, ρ_m, the surface density, σ, of the monolayers of small particles, and the torsion modulus, m. Before undertaking this task in earnest and with full attention to technical detail, however, it will be useful to survey, in a brief and intuitive manner, the whole question of the relevance of electrical units to the parameters of Maxwell's model; this brief overview will serve to establish a framework for the more complete analysis to follow.

The ratio of units, c, is in essence a measure of the relative strengths of electrostatic forces (electrical forces, exerted between electric charges by virtue of their electric charge) and electromagnetic forces (magnetic forces, exerted between electric currents, that is, moving electric charges, by virtue of their motion). The electrostatic force exerted between given, equal electric charges separated by unit distance is found by experiment to be much larger, in all of the usual systems of units (units of length on human dimensions, millimeters to meters, and time in seconds), than the electromagnetic force per unit length exerted when those same charges flow, in unit time, through parallel conductors separated by unit distance. (This comparison of forces actually gives the square of the ratio of units, as the force laws are quadratic in the charges, and an additional factor of 2 intervenes, owing to spatial integration of the primitive electromagnetic force law.[15]) Utilizing the experimental results of Weber and Kohlrausch, and converting, effectively, from their c_W [equation (5.2)] by dividing by $\sqrt{2}$, Maxwell obtained for the ratio between electrostatic and electromagnetic forces (i.e., the ratio between electromagnetic and electrostatic units) the following value:[16]

$$c = 310{,}740{,}000{,}000 \text{ millimeters/second} \tag{5.8}$$

In units of millimeters and seconds, then, electrostatic forces are larger by a factor of some 3×10^{11}. (Modernly, the same parameter is cited, typically, as $c = 2.997930 \pm 0.000003 \times 10^{10}$ cm/sec.[17])

In the molecular-vortex model, it is the density of the magnetoelectric medium, ρ_m, that controls the strength of magnetic forces: For a given electric current – that is, for given translational velocities of the small particles at the surfaces of the vortex cells – there will be given angular velocities of the vortices; the centrifugal forces engendered by these given rotational velocities of the vortices will then be proportional to the mass density of the vortex medium, ρ_m, and the associated magnetic forces will also be proportional to ρ_m. In a similar manner, it is the elastic constant m, the torsion modulus of the medium, that controls the strength of electrostatic forces: For a given electric charge – that is, for a given accumulation of the small particles – there will be a given pattern of elastic strain in the medium; the stresses in the medium associated with the given strains will then be proportional to the elastic constant m, and the associated electrical forces will also be proportional to m. One would then expect that the ratio between electrostatic and electromagnetic forces would depend on the ratio of the medium parameters determining the strengths of those forces – that is, the ratio m/ρ_m. Taking into account the fact that electrostatic and electromagnetic forces depend quadratically on electric charges and electric currents, respectively, we might expect the mathematical form of this dependence to involve a square root; neglecting numerical factors of order 1, we would then expect, on the basis of this intuitive argument, the following relationship between the parameters of the medium and the ratio of units:

$$c = \sqrt{\frac{m}{\rho_m}} \tag{5.9}$$

It should be noted that this argument as yet has nothing to do with light. What I have endeavored to show is that if Maxwell's theory of the magnetoelectric medium was to yield both electrostatic and electromagnetic forces, in the correct ratio, then it was necessary that the constants of the medium be set so as to satisfy something like equation (5.9), where c, the ratio of units, was to be specified on the basis of measurements of electrical and magnetic forces. Maxwell did in fact set the constants of the medium to satisfy equation (5.9) – we shall proceed presently to a discussion of how he actually went about that – but he did not, in "Physical Lines," make a point of isolating the relationship and arguing that it was purely electromagnetic, and not ad hoc to a theory of light. In a contemporaneous letter to William Thomson, however, Maxwell did point out that he had "deduced the relation between the elasticity and density of the cells" of the medium "by comparison with Weber's value of the statical measure of a unit of electrical current" – thus on purely electromagnetic grounds. Moreover, Maxwell explicitly denied that this had been ad hoc to a theory of light: He had "made out the equations in the country" – that is, at the family estate in Scotland – where he presumably did not have access to the latest numbers for the ratio of units and the velocity of light, and thus did not have "any suspicion of the nearness between the two values."[18] Our evaluation of that claim – to the effect that equation (5.9) arose naturally out of the model and not ad hoc to a hoped-for unification of electromagnetism and optics – must rest, however, on a careful analysis of how Maxwell actually did set the parameters ρ_m and m.

The setting of these parameters took place in the context of the progressive development of the theory of molecular vortices, as discussed in Chapter 3. Maxwell's primary goal in "Physical Lines" had been to construct a comprehensive mechanical theory of electromagnetic phenomena, and in order to achieve that kind of comprehensive coverage, he had employed the electromagnetic field equations to establish linkages between the electromagnetic variables; the mechanical part of the theory was then developed, in successive steps, to model those linkages and in that way yield a comprehensive theory, including mechanical analogues for all of the electromagnetic variables. The main chain of electromagnetic linkages – analyzed in detail in Chapter 3 – was as follows [designating successive electromagnetic linkages as (L1), (L2), etc.]:

(L1) Magnetic forces were related to magnetic fields through the magnetic stress tensor, as in equation (3.5).

(L2) Magnetic fields were related to electric currents through Ampère's law, as in schema (3.7).

(L3) Electric current was related to electric charge through Ampère's law with displacement current, along with the equation of continuity, as in schema (3.11) and equations (3.12)/(3.13) and (3.14).

(L4) Electric charge was related to the electric field through Coulomb's law, as in schema (3.15).

(L5) The electric field was related to electrical forces through the formula for electrical field energy and its gradient, as in equations (3.18) and (3.19).

What this provided, in sum, was a chain of linkage between magnetic forces at the beginning of (L1) and electrical forces at the end of (L5); that was what enabled Maxwell to integrate the ratio of units – specifying the relationship between magnetic and electrical forces – into his theory in a natural way.

Viewed in mechanical terms, the chain of linkage was as follows [designating the successive linkages, mechanically viewed, as (L1′), (L2′), etc.]:

(L1′) Magnetic forces were related to the rotational velocities of the vortices and the *density of the medium* through the mechanical stress tensor, as in equation (3.1).

(L2′) The rotational velocities of the vortices were related to the surface densities and flux densities of the small particles through the mechanical analogue of Ampère's law, equation (3.7a).

(L3′) The flux densities of the small particles were related to their accumulation points through the mechanical analogues of the displacement current and the equation of continuity – see equations (3.9), (3.11a), and (3.12)/(3.13).

(L4′) Accumulation points of the small particles were related to elas-

tic stresses and strains in the medium through the mechanical
analogue of Coulomb's law, equation (3.15a).

(L5′) Elastic stresses and strains in the medium and the *torsion modulus of the medium* were related to the elastic energy of the
medium, and thence to electrical forces, as in equations (3.18)
and (3.19).

It was through this chain of mechanical linkages that the *density of the
medium, ρ_m,* was related its *torsion modulus, m;* at the electromagnetic
level, the parallel linkage was from magnetic to electrical forces, the ratio
of which is measured by the ratio of units, c. The result of these parallel
linkages was to impose on the model a relationship between the mechanical parameters ρ_m and m and the electromagnetic constant c – as expressed in equation (5.9).

The way in which this relationship was realized in the progressive
development of the molecular-vortex model was as follows: In Part I of
"Physical Lines," Maxwell made an arbitrary choice (see Chapter 3) in
setting the value of the mass density of the medium, ρ_m, to be

$$\rho_m = \frac{1}{\pi}$$

That choice, propagated through the chain of mechanical linkages
(L1′)–(L5′), then forced the choice

$$m = \frac{c^2}{\pi}$$

for the torsion modulus of the medium, yielding the required relationship
as specified by equation (5.9). In order to see precisely how the initial
choice of ρ_m was propagated through the chain of mechanical linkages to
force a value for m consonant with equation (5.9), it will be useful to ask
how a change in the former would be propagated through the chain to
produce a change in the latter. Thus, as in equations (3.2′)/(3.3′), the
density could just as well have been set to

$$\rho_m = \frac{b}{\pi} \qquad (5.10)$$

where b is any positive (real) number. The parameter b then defines a
one-parameter family of possible models. According to linkage (L1′), the
mechanical stress tensor and the associated magnetic forces depend quadratically on the rotational velocities of the vortices and linearly on the
density of the medium, as in equation (3.1); thus, the magnitudes of

magnetic forces will be preserved invariate over the family of models defined by the parameter b if the rotational velocities are multiplied by $1/\sqrt{b}$, as in correspondences (3.2′) and (3.3′). In turn, according to linkage (L2′), the electric current – as represented by the flux density of the small particles – depends linearly on the rotational velocities of the vortices and also on the surface density of the small particles, σ, through equations (3.7a) and (3.9); in order to maintain the electric current invariate, therefore, σ must be multiplied by $\sqrt{b}$, so that Maxwell's

$$\sigma = \frac{1}{2\pi}$$

becomes

$$\sigma = \frac{\sqrt{b}}{2\pi}$$

as in equations (3.9) and (3.9′).

This choice for σ is in turn propagated through linkages (L3′), (L4′), and (L5′). As the electric currents (small-particle fluxes) have been maintained invariate, so also will the accumulations of electric charge (small particles) be maintained invariate by linkage (L3′), as in equation (3.11a). In linkage (L4′), however, the magnitude of an accumulation of the small particles depends linearly on both the surface density of the small particles and the elastic strains in the medium;[19] the elastic strains must therefore be multiplied by $1/\sqrt{b}$. Finally, in linkage (L5′), the elastic energy of the medium must be preserved invariate in order to preserve the magnitudes of electrical forces. As the energy depends linearly on the elastic constants and quadratically on the elastic strains – equations (3.17) and (3.18) – the elastic constants must be multiplied by b in order to preserve the energy invariate; Maxwell envisioned a medium that was characterized by a single elastic constant – namely, the torsion modulus m – so that his

$$m = \frac{c^2}{\pi}$$

becomes

$$m = \frac{bc^2}{\pi} \tag{5.11}$$

Finally, then, for a model characterized by any given value of the parameter b, the relationship between medium constants specified by equation

(5.9) will be preserved, as can be seen by taking the square root of the ratio of equations (5.11) and (5.10):

$$V = \sqrt{\frac{m}{\rho_m}} = \sqrt{\frac{bc^2/\pi}{b/\pi}} = c$$

That is, equation (5.9) will be satisfied in *the whole family of models* characterized by the parameter *b*, because the mass density ρ_m and the torsion modulus *m* depend in the same way on the parameter *b*, so that *b* cancels out in their ratio.

This exercise may serve to remove some of the mystery attendant upon Maxwell's choice of medium constants. The choice of ρ_m and *m* as respectively 1 and c^2, with associated factors of π, does have a look of the arbitrary or ad hoc about it. What now becomes clear is that although some measure of choice was involved, it extended only to the setting of a parameter multiplying both ρ_m and *m*, so that taking the ratio of the two served to eliminate any arbitrary element. Equation (5.9), therefore, emerged naturally and necessarily from the molecular-vortex model, rather than being imposed upon it ad hoc: The ratio of medium constants m/ρ_m was set in that model to get the correct ratio between electromagnetic and electrostatic forces, not to accommodate light in the model.

The model, then, was *determinative* of equation (5.9), independent of any considerations relating to optics, just as Maxwell claimed; the unification of electromagnetism and optics, however, flowed immediately from this relationship: As Maxwell put it, "by the ordinary method of investigation" – that is, from the mechanics of elastic media – "we know" that "the rate of propagation of transverse vibrations through the elastic medium of which the cells are composed" is given by[20]

$$V = \sqrt{\frac{m}{\rho_m}} \tag{5.7}$$

The relationship specified by equation (5.9) required that this ratio be equal to the ratio of units; making use of the data of Weber and Kohlrausch, as expressed in equation (5.8), and translating into English units, Maxwell concluded that the velocity *V* of transverse waves in the magnetoelectric medium would be given by

$$V = 193{,}088 \text{ miles per second}$$

It was only at this point in the argument that the velocity of light made its appearance; it will be worthwhile, at the risk of some slight repetition, to quote Maxwell in full:

> The velocity of light in air, as determined by M. Fizeau, is 70,843 leagues per second (25 leagues to a degree) which gives
>
> $$V = 314,858,000,000 \text{ millimetres}$$
> $$= 195,647 \text{ miles per second}$$
>
> The velocity of transverse undulations in our hypothetical medium, calculated from the electro-magnetic experiments of MM. Kohlrausch and Weber, agrees so exactly with the velocity of light calculated from the optical experiments of M. Fizeau, that we can scarcely avoid the inference that *light consists in the transverse undulations of the same medium which is the cause of electric and magnetic phenomena.*[21]

Maxwell made no explicit numerical estimate of the closeness of agreement between his calculation of the velocity of magnoelectric undulations and Fizeau's measurement of the speed of light; that was left as an exercise for the reader. Carrying out the calculation, we find that the two numbers differ by only 1.3 percent.[22] The agreement is thus quite close – in fact, too close, as the skeptic might point out: Given the various approximations that were made in the model calculations, why did Maxwell's calculation of the wave velocity turn out to be so close to the measured velocity of light? Were the mechanical calculations really that precise? Was Maxwell perhaps very lucky in his approximations? Or was there perhaps some residual adjustability in the mechanical calculations, allowing the wave velocity in the magnetoelectric medium – although basically determined by the measurements of Weber and Kohlrausch on the ratio of units – to be adjusted by factors of order 1, in order to agree with the measured velocity of light to within about 1 percent?

4   Approximations and adjustments

We have seen that parallel linkages of the main electromagnetic and mechanical variables – linkages (L1)–(L5) and (L1′)–(L5′), based principally on schemata (3.7), (3.11), and (3.15) – imposed a definite relationship between the ratio of electrical units and the mechanical parameters of the magnetoelectric medium, as in equation (5.9). The precision with which this last relationship could be established, however, could not be better than the precision of the mechanical calculations that were involved – including those that established equations (3.7a),

(3.11a), and (3.15a) – and these calculations were approximate rather than exact. Beyond this, we must inquire into other factors that might have affected equation (5.9), including calculational errors and adjustable parameters. All three of these – approximations, errors in calculation, and adjustable parameters – represent possible sources of factors of the order of 1 that could have intervened to alter the relationship between the mechanical parameters of the medium (and hence the velocity of wave propagation) and the ratio of electrical units.

The basic approximation that Maxwell made in his calculations leading to equations (3.7a), (3.11a), and (3.15a) involved approximating the shapes of the vortex cells to spheres in some parts of the calculations, while elsewhere making assumptions that required the shapes of the cells to deviate from perfect sphericity. Thus, in calculating the motions of the small particles, Maxwell assumed that each small particle would always be in contact with two adjacent vortex surfaces, as suggested visually in Figure 2.2.[23] This requires that the vortices be space-filling, except for the space taken up by the small particles; the use of the simple equation (5.7) for wave propagation also requires that the medium be space-filling.[24] The shapes of the space-filling cells would therefore have to deviate from perfect sphericity, and calculations assuming spherical cells thus had the status of approximations. Further, for the purpose of calculating the pressures in the medium giving rise to magnetic forces, Maxwell had considered the vortex cells as circular cylinders, which constituted another deviation from the spherical norm.[25]

Maxwell had been explicit about the approximative nature of his calculations, judging that the approximations he had made were probably reasonable: "The actual form of the cells probably does not differ from that of a sphere sufficiently to make much difference in the numerical result." Maxwell did not make any quantitative estimate of the possible deviation in numerical results, but it will be useful for our purposes to estimate a typical deviation. Thus, for example, comparing a right circular cylinder circumscribed about a sphere with that sphere as concerns surface area and volume – Maxwell used both of these characteristics in his calculations – we find that in both volume and surface area the cylinder differs from the sphere by a factor of 1.5; the deviation of a hexagonal cylinder from an inscribed sphere would be somewhat more than this, and the deviation of a dodecahedron somewhat less, so a factor of 1.5 would appear to be a characteristic error made in approximating to a sphere.[26]

Beyond these approximations of which he was aware, Maxwell made one clear error in his calculations, as was first pointed out by Pierre Duhem. The error stemmed from a problem in characterizing the elastic constants of the medium, so that, rather than Maxwell's m, this quantity divided by 2 should have been used in equation (5.7) to calculate the wave velocity.[27] Given this error of a factor of 2, as well as various approximative errors contributing multiplicative factors (or divisors) on the order of 1.5, it would not be surprising to have the result for the ratio m/ρ_m be off by a factor of 3 or 4, which would result in a deviation of up to a factor of 2 in the square root.

Given this situation, one would have to consider it a great stroke of luck for Maxwell to have come up with an answer for the velocity of wave propagation that was within 1 percent of any given number – in particular, the velocity of light. In fact, it was most probably not sheer luck, for Maxwell did have at least one adjustable parameter. An elastic medium is in general characterized by two constants, a bulk or compression modulus and a torsion or shear modulus, and depending on the ratio between these two moduli, Maxwell found that the relationship specified in equation (5.5) was in fact susceptible of some adjustment, within the range

$$\frac{c^2}{3} \leq \pi m \leq c^2$$

In support of the choice specified by equation (5.5) – which implies that the elasticity of the medium depends on pairwise central forces alone at the molecular level – Maxwell could make no argument; we find only the bald assertion that "it is probable that the substance of our cells is of [this] kind." I think it may not be unreasonable to surmise that this represented, for Maxwell, an adjustable parameter, with a range of a factor of 3, and he used this adjustability to achieve a particularly simple result, namely, that the velocity of propagation of waves in the medium would be given by the ratio of units, without additional numerical factors.[28]

Although the achievement of the simple result that the velocity of propagation of elastic waves in the magnetoelectric medium would be simply the ratio of electrical units may have been the result of a certain amount of ad hoc adjustment, the 1 percent agreement between the ratio of units and the measured velocity of light was not a matter subject to adjustment, as these were two experimentally measured parameters. Indeed, as Maxwell testified repeatedly and credibly, this very close agreement between the two experimentally measured parameters – the ratio of units and the velocity of light – came as a surprise to him: In addition to the letter to Thomson quoted in Section 3, there was a letter in a similar vein to Faraday, in which Maxwell wrote that "I worked out the formulae in the country before seeing Weber's number, which is in millimetres." That is, Maxwell knew the velocity of light in miles per second, whereas "Weber's number," the ratio of units as measured in the laboratory, was expressed in millimeters per second; Maxwell therefore would not have been aware of the close agreement between the two numbers until he converted one of them, which he did not do until he returned to London in the fall of 1861 and looked up the precise value of "Weber's number." (Furthermore, Weber's definition of the ratio of units differed from Maxwell's by a factor of $\sqrt{2}$, making the correspondence even harder to perceive by casual inspection.) Thus, I think we must believe Maxwell's testimony that he had no idea of the close agreement between the ratio of units and the velocity of light when he was developing the theory in the summer of 1861. What this means is that Maxwell was *not* adjusting his calculations to make the velocity of propagation of waves in the magnetoelectric medium come out close to the observed velocity of light. Whatever adjustment there may have been was directed, rather, toward obtaining a simple and elegant result entirely within the context of electromagnetic theory – namely, that the velocity of propagation of waves in the magnetoelectric medium was simply the ratio of units, with no accompanying numerical factors. That the ratio of units would turn out to be so close to the measured velocity of light was unanticipated – a pleasant surprise, which served to convince Maxwell that the unification of electromagnetism and optics was feasible. Maxwell also believed that this kind of result – 1 percent agreement with no fiddling – should convince Thomson and Faraday as well.[29]

Indeed, Maxwell believed that one of the main merits of his theory was to have focused attention on this close correspondence between the ratio of units and the velocity of light. This correspondence between a ratio of electrical forces measured in the laboratory and the observed velocity of light, in and of itself – independent of the details of Maxwell's theory – argued for a close connection between electromagnetism and optics: "This coincidence is not merely numerical . . . and I think we have now strong reason to believe, *whether my theory is a fact or not,* that the luminiferous and the electromagnetic medium are one," he wrote to Faraday.[30] The central conclusion in this connection, as Maxwell reiterated in his letter to Thomson, was that "Weber's number is really, as it appears to be, one half the velocity of light in millimetres per second"; that is, the very same number (sometimes multiplied or divided by 2 or $\sqrt{2}$, depending on definitions) appears in both electromagnetism and optics, and this in itself argues strongly for the unity of electromagnetic and optical phenomena.[31] There may have been some hyperbole in Maxwell's rhetoric – his confidence in the result was based on much more than mere numerology – but the contrast with Weber and Kirchhoff is nevertheless striking: For Maxwell – by his testimony – the numerical agreement (given the agreement in units) was in itself of significance; for Weber and Kirchhoff – by their testimony – it was not. Maxwell's rhetoric is particularly instructive in clarifying the sense in which the numerical agreement that so impressed Maxwell not only was independent of any ad hoc element in the theory of molecular vortices but also was substantially independent of the theory itself.

To sum up, there were two stages in Maxwell's argument to the conclusion that the magnetoelectric and luminiferous media were one and the same. First, there was the reasoning, based on the full intricacy of the molecular-vortex model, to the conclusion that the velocity V_M of waves in the magnetoelectric medium would be equal to the ratio of units:

$$V_M = c \qquad (5.10)$$

Second, there was the realization – based on comparing the observed value of the ratio of units c (as measured by Weber and Kohlrausch in their electrical laboratory) with the observed velocity of light V_L (as measured by Fizeau along a line of sight from Montmartre to the outskirts of Paris[32]) – that the velocity of light was very nearly equal to the ratio of units:

$$V_L \simeq c \qquad (5.11)$$

Putting these together, one obtains

$$V_M \simeq V_L \qquad (5.12)$$

from which Maxwell concluded that the magnetoelectric and luminiferous ethers must be one and the same. In arriving at the neat and simple numerical equality of equation (5.10) – by way of an intricate set of calculations, involving approximations, outright errors, and a choice of elastic parameters – I think one must judge that Maxwell was drawn forward by, and made accommodations toward, obtaining that simple result. That, however, was all done before Maxwell became aware of the surprisingly close agreement between the experimental results of Fizeau and those of Weber and Kohlrausch, as expressed in equation (5.11). The adjustments leading to equation (5.10) thus could not have been made ad hoc to obtaining equation (5.12), but rather were made on grounds of simplicity. Equation (5.11), which is completely independent of the theory of molecular vortices, was a startling discovery made by Maxwell in the library, and it involves no tinkering on his part whatsoever. Equation (5.12), finally, resulting from the composition of (5.10) and (5.11), receives a certain element of ad hoc character through the lineage of (5.10), reflecting, however, not an accommodation toward the final result but rather the use of a simplicity criterion in traversing a maze of difficult calculations.

The model, then, was determinative – to within factors of the order of 1 – in specifying the agreement of the calculated velocity of waves in the magnetoelectric medium with the observed velocity of light. At the level of factors of order 1, there had been errors, approximations, and adjustments in the calculations, but they turned out to be beside the point: The 1 percent agreement that was so impressive was between parameters established experimentally, by Fizeau and by Weber and Kohlrausch.

5 Other consequences

The calculation of a wave velocity that agreed closely with that of light was the central result underwriting Maxwell's unification of electromagnetism and optics on the basis of the molecular-vortex model. In addition to that, however, he was able to derive from the theory two further results involving connections between electromagnetic and optical phenomena; those results pertained to the refractive indices of dielectric media and the Faraday rotation.

In Part III of "Physical Lines," Maxwell derived a relationship between the dielectric constant of a medium and its refractive index, thus relating electromagnetic and optical properties. To see the physical basis for this in the molecular-vortex model, refer back to Figures 4.11 and 4.13, where the situation of the vortex cells within a charged capacitor ("Leyden jar") is depicted. The charge on the capacitor, for a given geometry, will be proportional to the polarization vector **P**, which gives rise to that charge; mechanically, however, **P** is proportional to the strain of the vortex material (where the proportionality constant depends on σ, the charge density of the small particles at the surfaces of the vortices). The electric charge on the plates, then, is proportional to the elastic strain in the medium. On the other hand, the voltage on the capacitor, again for given geometry, will be proportional to the electromotive force vector **E**, which gives rise to that voltage; mechanically, however, **E** is proportional to the stress on the vortex material (where the proportionality constant again depends on σ). The difference in potential between the plates, therefore, is proportional to the stress in the medium. As the relative dielectric constant, which Maxwell denoted D, is measured by the ratio of charge to voltage for a given geometry, it will correspond mechanically to the ratio of strain to stress – which, for Maxwell's single-constant medium, is given by the torsion modulus m. In optical terms, however, the torsion modulus m varies directly with the square of the wave velocity, as in equation (5.13). The relative dielectric constant D, therefore, will vary with the square of the relative wave velocity, or refractive index i:

$$D = \frac{i^2}{\mu} \qquad (5.13)$$

where the proportionality constant is the magnetic inductive power μ, as Maxwell determined by explicit calculation.[33]

This, in contrast to the result concerning wave velocity, was a prediction of a novel relationship, concerning which existing data could render no judgment. Predictive power was quite generally understood as providing an important criterion for the evaluation of theories,[34] and Maxwell appreciated the significance of this: As he wrote to Faraday, this kind of prediction of "numerical relations between optical, electric, and electromagnetic phenomena" was "capable of testing [the] theory"; Maxwell sought Faraday's help in assembling the data required for verification of the indicated relationship. Maxwell also queried Thomson as to whether or not the latter "kn[e]w any good measures of dielectric capacity of transparent substances," and in a letter to Cecil Monro, Maxwell indicated that he was planning some experiments of his own on dielectric constants. The matter was, however, not straightforward, and it was not until the 1870s that confirming evidence was obtained.[35] What was demonstrated immediately, then, was only the potential fruitfulness of the theory. Nor did the equations for the Faraday rotation that Maxwell derived from the theory provide immediate support (indeed, there were severe difficulties with these equations, which were never to be remedied within a strict Maxwellian framework).[36] These two further consequences relating to the unification of electromagnetism and optics that Maxwell derived from the molecular-vortex model thus served, initially at least, more to demonstrate the *potential* reach and unifying power of the theory than to offer direct and immediate empirical support. The central empirical mainstay of the theory remained the agreement between the calculated velocity of waves in the magnetoelectric medium and the measured velocity of light.

Conclusion

Within three years of the publication of "Physical Lines," Maxwell was to find a way to accomplish the unification of electromagnetism and optics in a more direct way, on the basis of the field equations, and it was that later work that was to be widely noted and approved. The primary significance of the initial unification on the basis of the molecular-vortex model, therefore, was as an intermediate step in the development of Maxwell's own thinking on the subject. That intermediate stage was important for the later work in two ways: First, it was the combination of conviction and doubt with which Maxwell regarded the initial unification, on the basis of the molecular-vortex model, that impelled him onward toward a new formulation of the connection between electromagnetism and optics. Second, it was the introduction of the displacement current into the equations, accomplished in the context of the molecular-vortex model, that made the later reformulation on the basis of the field equations possible. Maxwell's subsequent work on the displacement current and the electromagnetic theory of light will be taken up in Chapter 6. Although in that later work Maxwell was to move away from the molecular-vortex model, that mechanical model had, at a crucial juncture, been an engine of discovery for him, as concerned both the introduction of the displacement current and the unification of electromagnetism and optics.

6

Beyond molecular vortices

The theory of molecular vortices had constituted the focus of Maxwell's research program in electricity and magnetism in the late 1850s and early 1860s, and his two major innovations of that period – the introduction of the displacement current and the treatment of electromagnetism and optics within a single theoretical framework – grew out of the theory of molecular vortices and reflected that context in their initial formulations. In the course of Maxwell's elaboration of the molecular-vortex model, however, problems had accumulated, to the point that he had serious reservations concerning certain parts of the model. In addition, Maxwell's research program in the theory of heat and gases was, in the years around 1860 and thereafter, developing in such a way as to undermine support for the theory of molecular vortices in that area, which ha₁ been its original stronghold. Finally, and relatedly, Maxwell's general v ews on the use of mechanical models in science were developing in a new direction that involved less emphasis on specific and concrete models. All of these factors converged in encouraging Maxwell to begin a measured retreat from the molecular-vortex model.[1]

As part of this general retreat from the model, Maxwell took steps to free his signal innovations in electromagnetic theory from their original matrix in the theory of molecular vortices. The modification of Ampère's law and, more significantly, the incorporation of optics into electromagnetic theory defined new research programs, based on those innovations. Research programs defined in that way, however, contained the possibility of transcending the molecular-vortex connection and taking on lives of their own. Section 1 of this chapter deals with Maxwell's continuing research relating to the displacement current, in which he looked at the connections and implications of the new form of Ampère's law, independent of the model that had facilitated its formulation; he was led by this research to new formulations of the concept of displacement and its relationship to electric charge. Section 2 considers Maxwell's continuing program of research centering on the connection between electromagnetism and light, which included a relatively unsuccessful search for additional experimental support, as well as a supremely successful effort at recasting the theory, without reliance on the molecular-vortex model. Section 3 considers the reception of Maxwell's innovations, in the 1870s and thereafter, in the English-speaking world and in Europe. Section 4 considers in a synoptic manner the experimental validation of Maxwell's innovations by Heinrich Hertz and the subsequent amalgamation of British and Continental traditions, around the turn of the twentieth century, eventuating in the ultimate articulation of classical electromagnetic theory.

1   *Reformulating the displacement current*

In its first appearance, in the January 1862 installment of "Physical Lines," the displacement current had been quite thoroughly embedded in the molecular-vortex matrix. Indeed, Maxwell there treated the new form of Ampère's law basically as a mechanical calculation, whose function was to facilitate the extension of the molecular-vortex theory to embrace electrostatics; no mention was made of the electromagnetic significance of the new equation as a generalization of Ampère's law to the open circuit.[2] In private communication at that very time, however, Maxwell indicated that he was already thinking about the broader electromagnetic significance of his new equation, independent of the theory of molecular vortices. In December of 1861 – thus actually before the modified form of Ampère's law had appeared in print – Maxwell wrote to Henry R. Droop, an old Cambridge friend (third wrangler, just behind Maxwell's second on the 1854 mathematical tripos, and thus a worthy sounding board for Maxwell's new ideas), as follows: "I am trying to form an exact mathematical expression for all that is known about electromagnetism [i.e., electromagnetism proper, as covered by Ampère's law], *without the aid of hypothesis.*" Maxwell indicated that he was looking into the question of the magnetic effects of closed circuits as well as open circuits in the most general way, comparing formulas based on the charge-interaction point of view – involving forces at a distance between current elements – with formulas based on the field-primacy point of view – involving the calculation of a magnetic field quantity as an intermediary; his emphasis here was on the equations – on the "mathematical expressions" and the "mathematical reasons" – rather than on the question of a model underlying the equations.[3] Maxwell's efforts to divorce the new form of Ampère's law from the molecular-vortex model were to bear fruit within two years: In 1864 he presented a reformulated version that was completely independent of the theory of molecular vortices.

Further, in 1868, he clarified the electromagnetic significance of the new equation: This rendition of Ampère's law, finally, incorporated a composite solenoidal current to function as the source of the magnetic field.[4]

Little survives in the way of working papers from the period 1862–4 that would enable us to document Maxwell's precise path from the formulation of the displacement current within the context of the theory of molecular vortices, in 1862, to the reformulation, independent of molecular vortices, in 1864.[5] Comparison of the new formulation with the old, however, is suggestive of some of the considerations that may have been operative. The original definition of the displacement current had been directly tied to the mechanical details of the theory of molecular vortices. To see this, consider again the charging capacitor circuit, as in Figures 4.4–4.6. In the space between the capacitor plates, there are two contributions to the electric current – the **curl H** term and the displacement current – and these are distinguished mechanically, as arising respectively from the rotations and from the elastic deformations of the vortices. If one suppresses the mechanical aspects of the theory, as Maxwell was attempting to do during that period, there is no longer a basis for distinguishing two contributions to the current in the space between the capacitor plates. Maxwell's response to that was to alter his treatment of the displacement current significantly. In the new treatment of 1864 there was only one current in the space between the capacitor plates, namely, the displacement current. The displacement current had thus been promoted – if one may so put it – from a partial current in the dielectric to the whole current in the dielectric.

Relatedly, the direction of the displacement current had been reversed, as indicated by a reversal of sign [as compared with equations (4.11)/(4.14)] in the relationship between electric field and displacement:

$$P = kf$$
$$Q = kg \qquad (6.1)$$
$$R = kh$$

The components P, Q, R here again represent the "electromotive force" and can be rendered vectorially as the electric field **E** (see Chapter 4, Section 2). Once again, f, g, h is the "electric displacement," which we rendered before as a polarization vector **P** (Chapter 4, Sections 1 and 4). As redefined here, f, g, h is more directly assimilable to the displacement vector **D** of the standard account, but as will become evident from Maxwell's treatment of the time derivative of f, g, h, this vector must differ from the modern $\mathbf{D}$ by a factor of 4π (given unrationalized electromagnetic units for the standard account, as in Chapter 4, Section 1); to reflect this mnemonically, f, g, h will be rendered as a vector $\mathbf{D}_M$ (M for Maxwell), where

$$\mathbf{D}_M = \frac{1}{4\pi} \mathbf{D}$$

The constant k was defined as "the ratio of the electromotive force to the electric displacement," which is basically the inverse of the modern dielectric constant ϵ, but with a factor of 4π again intervening.[6] Equation (6.1) thus becomes

$$E = \frac{4\pi}{\epsilon} \mathbf{D}_M$$

With displacement defined in this way – pointing in the same direction as the electric field – its time derivative, the displacement current, can be added to the true or conduction current to form a closed loop, as in the standard account (Figures 4.1–4.3). As Maxwell put it, "the variations of the electrical displacement must be added to the currents p, q, r to get the total motion of electricity, which we may call p', q', r'":

$$p' = p + \frac{df}{dt}$$

$$q' = q + \frac{dg}{dt}$$

$$r' = r + \frac{dh}{dt}$$

where p', q', r' is also called the "total current"; the equation becomes, in mnemonic vector notation.

$$\mathbf{J}' = \mathbf{J} + \frac{\partial \mathbf{D}_M}{\partial t} \tag{6.2}$$

In this equation, as in the standard account, the current $\mathbf{J}$ and the displacement current exist in distinct spatial locations, and the addition gives rise to a closed loop, as in Figure 4.3b: "According to this view, the current produced in discharging a condenser is a complete circuit."[7]

Having constructed a solenoidal composite current $\mathbf{J}'$, Maxwell was able to equate $4\pi\mathbf{J}'$ to the solenoidal **curl H** term in Ampère's law, thus achieving consistency for the open circuit in the manner described in the standard account:

$$\text{curl } \mathbf{H} = 4\pi\mathbf{J}' \tag{6.3}$$

where the rendering of Maxwell's component equations is unproblematic. Substituting equation (6.2) into equation (6.3), one obtains

$$\text{curl } \mathbf{H} = 4\pi\left(\mathbf{J} + \frac{\partial \mathbf{D}_M}{\partial t} \right) \tag{6.4}$$

which, given that $\mathbf{D}_M = (1/4\pi)\mathbf{D}$, becomes exactly equation (4.2″) of the standard account.[8] Thus reformulated, the modified Ampère's law – with the displacement current regarded as a source of the magnetic field – could be used along with Faraday's law of electromagnetic induction to derive the propagation of electromagnetic waves with the velocity c; Maxwell accomplished that in a straightforward and eminently successful manner, as discussed in Section 2 of this chapter.

Not quite so straightforward and successful was Maxwell's attempt to adjust the other parts of his electromagnetic theory to accommodate the new formulation of Ampère's law and the displacement current. Consider again the paradigmatic case of the charging capacitor circuit: In Maxwell's original formulation, as represented in Figures 4.4 and 4.5, the displacement current was a reverse current (as compared with the current in the wire), associated with a growing reverse polarization. This dovetailed nicely with Maxwell's view of the nature of electric charge as emergent from field processes: The charges on the plates could be viewed as resulting from this reverse polarization (Figures 4.6 and 4.7). Maxwell's original equation for the relationship between displacement and electric charge embodied this perspective for the general case:

$$\rho = -\text{div } \mathbf{P} \tag{4.15}$$

In the 1864 formulation, however, the displacement current was to be seen as a forward current, associated with a growing forward polarization, and that would seem to require that the displacement $\mathbf{D}_M$ represent a polarization vector pointing away from the accumulation of positive charge, as in the standard account (Figures 4.1–4.3). Mathematically, the reformulated Ampère's law,

$$\text{curl } \mathbf{H} = 4\pi\left(\mathbf{J} + \frac{\partial \mathbf{D}_M}{\partial t} \right) \tag{6.4}$$

taken together with the equation of continuity,

$$\frac{\partial \rho}{\partial t} + \text{div } \mathbf{J} = 0 \tag{6.5}$$

would require

$$\frac{\partial \rho}{\partial t} = -\text{div } \mathbf{J}$$

$$= -\text{div}\left(\frac{1}{4\pi} \text{ curl } \mathbf{H} - \frac{\partial \mathbf{D}_M}{\partial t}\right)$$

$$= +\text{div } \frac{\partial \mathbf{D}_M}{\partial t}$$

or

$$\rho = +\text{div } \mathbf{D}_M \tag{6.6}$$

This result, unfortunately, differs in sign from the one required by Maxwell's original theory of electric charge.

Maxwell's response to this in the paper of 1864 was to write down, for the "Equation of Free Electricity," not the relationship required mathematically by his new formulation of the displacement current, equation (6.6), but rather

$$e + \frac{df}{dx} + \frac{dg}{dy} + \frac{dh}{dz} = 0 \tag{6.7}$$

or, in mnemonic rendering,

$$\rho = -\text{div } \mathbf{D}_M \tag{6.8}$$

Writing the equation in this way preserved the field-primacy perspective on electric charge, as Maxwell pointed out explicitly: "Let e [ρ] represent the quantity of free electricity contained in unit of volume at any part of the field, [which] arises from the electrification of the different parts of the field not neutralizing each other."[9] The price of this, however, was mathematical consistency: When Maxwell, in Part III of the paper, assembled the "General Equations of the Electromagnetic Field," as twenty equations in twenty variables (he counted each component equation, included potentials, and also counted constitutive equations and the equation of continuity), his equations A and C [together specifying the new form of Ampère's law as in equation (6.4)] were inconsistent with his equations G and H [respectively the equation of continuity and the divergence equation for $\mathbf{D}_M$, as in equations (6.5) and (6.8)].[10]

The difficulty that Maxwell faced was fundamental: He was fully committed to the new interpretation of the displacement current and Ampère's law, which required that displacement point *away from* concentrations of positive charge; he was also fully committed to the view of electric charge as emergent from the field, which appeared to require that the displacement point *toward* concentrations of positive charge.[11] That dilemma remained unresolved through the 1860s, but Maxwell was finally able to present a solution in the *Treatise on Electricity and Magnetism* (1873). There, Maxwell presented a complete and consistent set of equations, expressed in quaternion notation as well as component form, and corresponding very closely to the modern equations in the standard account. In particular, the algebraic sign in the divergence equation for the displacement D_M was positive, as in equation (6.6), making it consistent with the new formulation of the displacement current. Interestingly enough, in adopting equation (6.6), Maxwell did not have to give up completely his notion of electric charge as emergent from the field. It is true that the charge ρ could no longer be regarded as the usual Poisson bound charge arising from volume inhomogeneities or terminating surfaces of the D_M field, as, for example, in Figure 4.7. Rather, what was required, in order to make sense of equation (6.6) in the context of the field-primacy approach, was a radical reinterpretation of the meaning of the vector field D_M and its relationship to electric charge.[12]

Central to the new interpretation was a redefinition of the relationship between "electricity" and "charge." In the traditional, charge-interaction approach, "electricity" was a mobile fluid (or two mobile fluids) consisting of microscopic electric charges; a net accumulation of microscopic charges, or particles of electricity, of one sign would give rise to a macroscopic charge of that sign. In the molecular-vortex model, Maxwell had not achieved a complete emancipation from that approach: Even though, at the macroscopic level, charge was emergent from the field, still, at the microscopic level, the idle-wheel particles had played the role of primitive, mobile electric charges, and their accumulation points constituted macroscopic charges; in that, the model of "Physical Lines" made its compromise with the charge-interaction approach, and that was in fact one of the reasons for Maxwell's dissatisfaction with the model. Maxwell's new theory, however, as presented in the *Treatise,* was to be uncompromising in its rejection of the charge-interaction approach.[13]

In the new theory, then, there were to be *no primitive charges or microscopic charge carriers.* "Electricity" was to be viewed as an abstract, mobile, incompressible fluid that did not consist of electric charges and that did not have accumulation points. "Charge" thus could not be regarded as an accumulation of this fluid, both because the fluid did not consist of electric charges in the first place and because the fluid could not accumulate. In the charging capacitor circuit, there would be a solenoidal flow of "electricity," corresponding to Maxwell's composite solenoidal or "total" current, comprising the conduction current in the wire and the displacement current in the dielectric, as in Figure 4.3b. Electric "charge," finally, would arise from the effects that the flow of "electricity" would have on the medium through which it flowed. At some point in the charging process, let the vector $\mathbf{D}_M$, the "displacement of electricity," evaluated at a spatial point in the dielectric medium, be given by "the whole quantity of electricity" that has flowed, or been "displaced," through a unit of area at that point during the charging process (this amounts to the time integral of the displacement current). The dielectric medium remembers, as it were, how much "electricity" has been displaced through it, and the vector $\mathbf{D}_M$ measures a "state of constraint," or "electric stress," that has been produced by the flow of "electricity" through the medium since the beginning of the charging process. In a conducting medium, on the other hand, there is no long-term memory of the integrated flow of electricity; in such a medium, "the electric stress is relaxed," because "the conductivity of the medium allows it to decay," and so no $\mathbf{D}_M$ vector can be sustained in a conducting medium. At the boundary where the conducting capacitor plates meet the dielectric medium, therefore, there is a discontinuity in the $\mathbf{D}_M$ vector, and *it is this discontinuity in the $\mathbf{D}_M$ vector that is manifested as electric charge.* Mathematically, the surface charge at the boundary of the dielectric is given by the inner product of the $\mathbf{D}_M$ vector with the normal to the surface *directed inward* – as required by the positive sign in equation (6.6) – and this gives a consistent value for the electric charge.[14]

With that, Maxwell had achieved a consistent mathematical rendition of Faraday's picture of the relationship between electric lines of force and electric charge, namely, that electric charge is the manifestation of endpoints of lines of force, with the $\mathbf{D}_M$ field representing the lines of force mathematically. Maxwell's account was thus completely faithful to the field-primacy point of view, because electric charge had no independent existence, at the macroscopic or microscopic level, apart from the line of force or $\mathbf{D}_M$ field; the field was indeed primary, in that there were no primitive electric charges, and all charge was emergent from the field. In addition, this account of the nature of electric charge dovetailed nicely with the new treatment of Ampère's law and the displacement current, achieving mathematical consistency there, and also providing a physical interpretation of the composite solenoidal current, as representing the flow of "electricity" – something that was not itself electric charge, but whose passage through differing media gave rise to lines of stress or constraint whose endpoints were manifested as electric charge. The achievement of a complete mathematical and physical integration of the displacement current into electromagnetic theory, in the context of a physical picture that was completely faithful to the field-primacy point of view, was a fitting achievement for Maxwell's monumental *Treatise on Electricity and Magnetism* and an appropriate capstone for his research program in that area, and that was Maxwell's last word on the subject.[15] Others, however, were not quite so satisfied with that as a final resting place for the theory, as will be explored in Section 3.

2   *New foundations for the theory of light*
        Even before the publication of Part III of "Physical Lines" in January 1862, Maxwell had communicated his new result concerning the connection between electromagnetism and optics to Faraday, Thomson, and Cecil Monro. Monro was an old Cambridge friend with whom Maxwell corresponded regularly, and Monro's response is revealing:

> The coincidence between the observed velocity of light and your calculated velocity of a transverse vibration in your medium seems a brilliant result. But I must say I think a few such results are wanted before you can get people to think that, every time an electric current is produced, a little file of particles is squeezed along between rows of wheels. . . . I admit [however] that the possibility of convincing the public is not the question.

Maxwell, however, was in fact quite concerned with the question of "convincing the public": He saw himself as an advocate for the British field-primacy approach to electricity and magnetism, as against the Continental charge-interaction approach, and that advocacy was central to his research program and his publications in electricity and magnetism. Thus, given the admittedly farfetched nature of certain aspects of the theory of molecular vortices – and the picture of "little files[s] of particles [being] squeezed along between rows of wheels" was not the least of those – it is not at all surprising that he believed he could improve his opportunities for "convincing the public" by divorcing the treatment of light from the theory of molecular vortices. As he was to write to another friend when he had accomplished that, "I have . . . cleared the electromagnetic theory of light from all unwarrantable assumption." This is not to say that Maxwell had lost all faith in the theory of molecular vortices (see again Chapter 2); it was simply that developing and support-ing his theory of light had become an independent goal, and the theory of molecular vortices had come to seem perhaps more a hindrance than a help in that respect. The effort to free his theory of light from "unwarrant-able assumption" was to bear fruit in 1864, when Maxwell was able to present a true electromagnetic theory of light, proceeding directly from the field equations.[16]

There were other ways in which Maxwell pursued his research program centered on the theory of light in the period after the publication of "Physical Lines." The ratio of units played a central role in the theory, and Maxwell became involved in a program of experimentation directed toward more precise measurement and specification of electric quantities, electrical standards, and electrical units, including the ratio of units c; connected with this were theoretical investigations into the question of electrical units and dimensional analysis in general.[17] Finally, Maxwell continued to make efforts to obtain data on the dielectric constants of optically transparent media, in order to test his prediction of a relationship between dielectric constant and refractive index. In the event, however, it was the effort at reformulating the theory of light independently of the theory of molecular vortices that bore fruit first.

The new formulation of the displacement current, as we have seen, exhibited a composite solenoidal current – the sum of conduction and displacement currents [equation (6.6)] – as the source of the magnetic field, with the implication that a changing D_M or E field would give rise to an H field; coupled with Faraday's law of electromagnetic induction, which says that a changing B or H field will give rise in turn to an E field, that provided the physical basis for leapfrogging E and H fields constitut-ing electromagnetic waves. Mathematically, Maxwell carried out the der-ivation of transverse electromagnetic waves, propagating with velocity c, in two ways. First, in 1864, he carried out the derivation in full generality utilizing the vector and scalar potentials, in a calculation comparable to that employed in a modern derivation of the inhomogeneous wave equa-tion.[18] Second, in 1868, he gave a simplified and more physically sugges-tive argument, intended to convey the crux of the matter in its "simplest form."

In the simplified account of 1868, Maxwell supposes that all elec-tromagnetic quantities are functions of the z coordinate only; he con-siders, in the final analysis, only the air or vacuum case; and he postulates a magnetic field in the y direction – designated β in his notation, or H_y in mnemonic notation. According to Faraday's law of electromagnetic induction in differential form, a variation with time of H_y will give rise to an electric field in the x direction – designated P by Maxwell, or E_x in mnemonic notation – according to the following equation:

$$\frac{\partial E_x}{\partial z} = -\frac{\partial H_y}{\partial t}$$

Letting c be the ratio of units (Maxwell's V or $\sqrt{k/4\pi}$ here), the new form of Ampere's law says that a time variation of E_x will in turn give rise to a further H_y as follows:

$$\frac{\partial H_y}{\partial z} = -\frac{1}{c^2}\frac{\partial E_x}{\partial t}$$

Differentiating these two equations respectively with respect to t and z, and eliminating the mixed derivative of E_x, Maxwell obtained a wave equation,

$$\frac{\partial^2 H_y}{\partial t^2} = \frac{1}{c^2}\frac{\partial^2 H_y}{\partial z^2}$$

with solutions in the form of propagating waves of arbitrary form, with magnetic and associated electrical disturbances transverse to the direction of propagation, proceeding in the $+z$ or $-z$ direction at velocity c.[19]

This derivation of propagating transverse waves (or the more abstract and general derivation involving the potentials) involved no mechanical hypotheses, but only the electromagnetic equations. Furthermore, as Maxwell pointed out in a most trenchant manner, the argument to this point had been purely electromagnetic, including the measurement of c:

> The value of [c] was determined by measuring the electromotive force with which a condenser of known capacity was charged, and then discharging the condenser through a galvanometer, so as to measure the quantity of electricity in it in electromagnetic measure. The only use made of light in the experiment was to see the instruments[!]

The result of the experiment – for which Weber and Kohlrausch were cited once again – was

$$c = 310{,}740{,}000 \text{ meters per second}$$

On the other hand, the velocity of light in air – as recently measured and reported by Lèon Foucault in 1862 – was

$$V = 298{,}000{,}000 \text{ [meters per second]}$$

That result was purely optical in origin:

> The value of V found by M. Foucault was obtained by determining the angle through which a revolving mirror turned, while the light reflected from it went and returned along a measured course. No use whatever was made of electricity or magnetism.

The agreement of the velocity of propagation of electromagnetic waves – as calculated *without approximation* on the basis of purely electromagnetic equations and constants – with the velocity of light as measured by purely optical means constituted persuasive evidence for the electromagnetic nature of light:

> The agreement of the results seems to shew that light and magnetism are affections of the same substance, and that light is an electromagnetic disturbance propagated through the field according to electromagnetic laws.

This, finally, was the *locus classicus* for what may properly be termed the electromagnetic theory of light, as completely freed from the specifics of any mechanical ether theory – although Maxwell did continue to refer, at least in passing, to the "substance" that was the basis of both electromagnetic and optical phenomena.[20]

The *Treatise on Electricity and Magnetism* did not bring any further reformulation of the electromagnetic theory of light: The mathematical formulation of Ampère's law – if not the physical interpretation – remained the same, and the derivation of a wave equation from the combination of Ampère's law and Faraday's law also remained the same. There had been some progress, however, in ancillary parts of Maxwell's research program dealing with the electromagnetic theory of light. The arduous and exact experiments on the ratio of units had been completed, and William Thomson had reported some similar experiments; together with the experiments of Weber and Kohlrausch, these furnished three independent measurements of the ratio of units, which Maxwell was able to compare with three separate measurements of the velocity of light. The data were presented in tabular form as follows:[21]

Velocity of Light (mètres per second)		Ratio of Electric Units (mètres per second)	
Fizeau	314000000	Weber	310740000
Aberration, &c., and Sun's Parallax	308000000	Maxwell	288000000
Foucault	298360000	Thomson	282000000

The conclusion from all of this was that further experimentation had not eventuated in decisive verification of the electromagnetic theory of light (in the form of closer agreement between the ratio of units and the velocity of light) as Maxwell had hoped: The principal comparison that Maxwell had made in 1862 was between the measurements of Weber and Fizeau (the latter given there, to more significant figures, as 314,858,000,000 mm/sec), and those were a bit over 1 percent apart. The presumably improved measurements of the ratio of units by Maxwell and Thomson were even further from Fizeau's number (respectively 8 and 10 percent below), and even the lowest of the measurements of the velocity of light (Foucault's newer value, which Maxwell had favored in "Dynamical Theory") was larger than the results of Maxwell and Thomson by 4 and 6 percent, respectively. Maxwell's conclusion was a tempered one:

> It is manifest that the velocity of light and the ratio of the units are quantities of the same order of magnitude. Neither of them can be said to be determined as yet with such a degree of accuracy as to enable us to assert that the one is greater or less than the other. It is to be hoped that, by further experiment, the relation between the magnitudes of the two quantities may be more accurately determined.
>
> In the meantime our theory, which asserts that these two quantities are equal, and assigns a physical reason for this equality, is certainly not contradicted by the comparison of these results such as they are.[22]

Nor did the eagerly awaited data on the relationship between dielectric constant and refractive index prove to be any more decisive. Equation (5.13) predicted, for the refractive index in a nonmagnetic material,

$$i = \sqrt{D}$$

where i is the refractive index and D is the relative dielectric constant. The "only dielectric," however, for which the latter had been measured "with sufficient accuracy" was paraffin; the reported value from that measurement in 1871 was $D = 1.975$. The prediction of the theory for the refractive index of paraffin, therefore, was

$$i = \sqrt{1.975} = 1.405$$

Extrapolating from measured values of the refractive index of paraffin to the limit of long wavelength – for comparability with the static value of the dielectric constant – Maxwell arrived at a value of 1.422 for the refractive index. Comparing the two numbers, Maxwell arrived at a conclusion similar to his earlier judgment concerning the ratio of units and the velocity of light: The theory seemed generally on target, but the numbers were a bit off – the discrepancy was "greater than can be accounted for by errors of observation" – and thus did not furnish a confirmation of the most decisive kind.[23]

An account of the Faraday rotation had been another anticipated payoff of the electromagnetic theory of light, and Maxwell pursued that issue further in the *Treatise* as well, but once again, no substantial further support for the theory could be garnered. In the first place, in a theory that denied the existence of microscopic charge carriers, the kinds of explanations for the Faraday rotation that were being broached at the time by the Continental theorists (and have survived in modern theory) were unavailable to Maxwell, and his only recourse was to revert to the molecular-vortex picture. Though he still believed in molecular vortices in principle, Maxwell had by that point abandoned the attempt to specify a detailed molecular-vortex mechanism, and he had tried to avoid molecular-vortex considerations in the *Treatise* insofar as possible. As concerned the Faraday effect, however, Maxwell needed the vortex mechanism, and he also found that he had to make some very specific assumptions concerning the mechanics of the vortices. His conclusions on the Faraday rotation, therefore, had to be presented with the following caveat:

> The theory proposed in the preceding pages is evidently of a
> provisional kind, resting as it does on unproved hypotheses re-
> lating to the nature of molecular vortices, and the mode in
> which they are affected by the displacement of the medium.

Furthermore, even with the service of "unproved hypotheses" concerning the vortices, Maxwell remained unable to account in a consistent manner for some of the basic features of the Faraday rotation, such as the difference in sense of rotation between paramagnetic and diamagnetic materials. Maxwell had to conclude that knowledge of the molecular constitution of matter was not yet sufficiently developed to supply a proper account of the interaction of light and matter, and thus his account of the Faraday rotation, such as it was, could not provide any substantive support for the electromagnetic theory of light.[24]

Thus, by the time of the publication of the *Treatise* in 1873 – and indeed at the time of Maxwell's death in 1879 – his attempts to provide further experimental support for the electromagnetic theory of light had proved unavailing: Efforts to extend and improve agreement between theory and experiment, as concerned the ratio of units, the relationship between dielectric constant and index of refraction, and the Faraday rotation, had all failed to provide further support for the theory. The main accomplishment of Maxwell's research program in this area after "Physical Lines" had been the development of a true electromagnetic theory of light, proceeding directly from the field equations, and thus completely independent of the theory of molecular vortices.

3   *After Maxwell: the two traditions*

The electromagnetic theory of light, the displacement current, and their field-theoretic context became matters of interest and concern throughout the international physics community in the 1870s and thereafter. Maxwell's innovations were, however, differently regarded and differently developed in the English-speaking world and on the Continent. The primary vehicle for the dissemination of Maxwellian electromagnetic theory in Britain and the United States was the *Treatise on Electricity and Magnetism*, published in 1873 (second and third editions in 1881 and 1891). The *Treatise*, which was the climactic expression of the field-primacy approach to electromagnetic theory, formed the basis for a Maxwellian tradition in the English-speaking countries that was quite strictly faithful to that approach. On the Continent, it was Hermann von Helmholtz's paper "On the Equations of Motion of Electricity in Conducting Media at Rest," published in 1870, that constituted, initially, the main conduit for Maxwellian ideas; Helmholtz's paper attempted to assimilate Maxwell's signal innovations to the charge-interaction approach, thus providing the basis for an ongoing Continental tradition that represented a compromise position.[25]

The reception of Maxwell's innovations in Great Britain was, as might be expected, quite positive. The nexus for the development and dissemination of Maxwellian ideas in Great Britain was Cambridge University, and especially the Cavendish Laboratory. Maxwell had accepted the post of Cavendish Professor in 1871, had directed the planning and construction of the laboratory building, and then had presided over the institution from the time it admitted its first student in 1874 until Maxwell's death in 1879. During his tenure, Maxwell directed the work of some fifty students, and the central thematic element of the laboratory work was Maxwell's electromagnetic theory; included was substantial work on electrical standards that was relevant both to electrical technology and to the electromagnetic theory of light. The basic textbook from which the students learned their electromagnetic theory was the *Treatise*, and they came out well tutored in field theory, the electromagnetic theory of light, and the arcana of Maxwell's view of electric charge and displacement. The first generation of Cavendish students included William Garnett, Richard Glazebrook, William Hicks, William Davidson Niven, John Henry Poynting, and Arthur Schuster, among others. Other Cambridge graduates, through the 1880s, who made significant contributions to the development of the Maxwellian tradition in the later nineteenth century included Horace Lamb, Joseph Larmor, John William Strutt (Lord Rayleigh, second director of the Cavendish), and J. J. Thomson (third director). Other British followers in the later nineteenth century included George Fitzgerald, Oliver Heaviside, Oliver Lodge, and Sylvanus P. Thomson, and in the United States there were Edwin Hall and Henry Rowland. These and other less well known individuals formed an active and interacting group of thirty to forty "Maxwellians"; the group flourished through the last quarter of the nineteenth century, developing, extending, and testing Maxwell's electromagnetic theory, including his signal innovations, all in a context of quite strict fidelity to the basic tenets of the theory as laid down in the *Treatise*.[26]

Not every British scientist was so enthusiastic about Maxwell's innovations in electromagnetic theory. Most saliently, William Thomson – source of many of the ideas that were seminal for Maxwell – who had become Lord Kelvin, the most honored and well known scientist in Britain, took exception to both the displacement current and the electromagnetic theory of light. As concerned the former, Kelvin believed that it had not been demonstrated that there was any "conceivable hypothesis," or physical model, that could supply a conceptual basis for the displacement current. It would appear that Kelvin was most perplexed about situations in which conduction currents and displacement currents might coexist; Maxwell's reformulated theory, cut loose from any definite mechanical model, was quite sketchy concerning the nature of electric current, in particular as might apply to the case of coexisting conduction and displacement currents. As for any mathematical necessity for the displacement current, Kelvin took a dyspeptic view of "anyone trying to evolve out of his inner consciousness," whether by mathematical means or other, "a theory of the mutual force and induction between incomplete circuits"; experiment would have to be the final arbiter, and in the meantime Kelvin had some suggestions of his own concerning the effects of open circuits. (That the displacement current gave a neat solution for the case of the quasi-statically charging capacitor circuit would not have had particular force for Kelvin, because he was more concerned with radiative situations in the presence of metallic conductors, as in telegraphy, where the mathematical situation – especially if the potentials are used – is not so straightforward.[27])

Related to Kelvin's stance on the displacement current were his views on the electromagnetic theory of light. For Maxwell, one of the most important payoffs of the reformulated, disembodied displacement current was a parsimonious and truly electromagnetic theory of light; the successes of the electromagnetic theory of light in turn furnished support for the displacement current. Kelvin, however, did not see much of value in a purely electromagnetic theory of light; he was a great believer in a universal ether and thus was in sympathy with the enterprise of grounding electromagnetic and optical phenomena in one ether, as Maxwell had done in "Physical Lines." Kelvin, however, had no sympathy with Maxwell's subsequent attempt to liberate optical theory from a concrete mechanical basis by reducing optics directly to electricity and magnetism, rather than reducing both to the mechanics of one ether. Indeed, it seemed to Kelvin "rather a backward step from an absolutely definite notion that is put before us by Fresnel and his followers to take up the so-called Electro-magnetic theory of light in the way it has been taken up by several writers of late." As concerned both the displacement current and the electromagnetic theory of light, then, Thomson was critical of the main thrust of Maxwell's research subsequent to 1862, which was to liberate the theory from a concrete mechanical picture of the ether; Maxwell had turned away – in practice if not in principle – from the tenets of mechanical reductionism that had informed Thomson's teaching and Maxwell's learning from him in the 1850s, and Thomson did not forgive him for that.[28] Moreover, there was in the final decades of the nineteenth century a quite general trend away from mechanistic foundations in physics – yielding, according to Thomson, a position of "mere nihilism, having no part or lot in Natural Philosophy." Thomson's rejection of the electromagnetic theory of light was part of his taking a firm stance against this whole trend: "I never satisfy myself until I can make a mechanical model of a thing. If I can make a mechanical model I can understand it. As long as I cannot make a mechanical model all the way through I cannot understand; and that is why I cannot get the electro-magnetic theory."[29]

The exaggerated mechanistic commitment that Kelvin carried with him into the closing years of the nineteenth century was that of a pioneer of the new, energy-based mechanism of midcentury who had lived on for another fifty years to become a holdout for what were then seen as superannuated doctrines. Certain aspects of Kelvin's opposition to Maxwell's theory were, however, widely shared, on the Continent especially, if not so much among the British Maxwellians. First, there was the question of Maxwell's account of electric charge as arising from displacement, which was embraced by the Maxwellians, but was more skeptically viewed by both Kelvin and the Continentals. In addition, the problem of the interaction of light and matter – as, for example, in the Faraday rotation – had become central to research in optics in the later nineteenth century, and Maxwell's theory turned out to be not very helpful in that regard; the somewhat doleful record of Maxwell's successive attacks on the Faraday rotation had been a harbinger of that. As it turned out, the two problems – the nature of electric charge and the interaction of light and matter – were interconnected, and their resolution required a further recasting of the theory, which was begun by the Continental electrical theorists.[30]

Hermann von Helmholtz's seminal paper of 1870 sought to assimilate Maxwell's innovations, especially the electromagnetic theory of light, to the charge-interaction tradition. The proximate background for Helmholtz's paper was Maxwell's presentation in "Dynamical Theory"; there was also, however, considerable background for Helmholtz's initiative from within the charge-interaction tradition. In the first place, the possibility of an etherial connection for electricity and the further possibility of a connection with the luminiferous ether were not at all foreign to the charge-interaction tradition, from Ampère forward; second, the Continental researchers had all along paid careful attention to Faraday's experimental research program, and although their main concern was to assimilate Faraday's experimental novelties to their own theoretical approach, they could not help but be aware of the outlines of Faraday's interpretive framework. Moreover, there were specific developments within the charge-interaction tradition that prepared the way for Helmholtz's embrace of Maxwell's innovations. First, there had been Kirchhoff's analysis in 1857, along strict Weberian lines, of the propagation of waves of electric current in conducting media, at velocities comparable to that of light. Although Kirchhoff and his Continental readers had not been inclined to make a direct leap from that result to an electromagnetic theory of light, the analysis was nevertheless suggestive concerning the propagation of electromagnetic effects, and when Helmholtz came to undertake the project of assimilating Maxwell's innovations to the Continental tradition, he was able to do that by means of a generalized and updated version of Kirchhoff's original analysis.[31]

Further Continental background relevant to Helmholtz's embrace of Maxwell's ideas was furnished by a long history of discussions of the shortcomings of Weber's law of electrical action with respect to the principle of conservation of energy. Weber's use of velocity-dependent forces and potentials was in apparent conflict with the foundations of the law of conservation of energy, as stated by Helmholtz in his classic paper of 1847, and that furnished a basis for continuing critiques of Weber's law and searches for alternatives by Helmholtz and others.[32] One such alternative involved the notion of the propagated or retarded potential, which was discussed in private communication by Karl Friedrich Gauss and Bernhard Riemann in the 1850s and was discussed in print in the 1860s by Ludwig Lorenz, Riemann, and Carl Neumann.[33] The idea was that the interaction between two charges, for example, was not instantaneous, but rather took time to propagate from one charge to the other, with the propagation perhaps involving an intervening medium. This notion had precedent within the charge-interaction tradition: From Ampère on, the idea of propagation of interaction through a medium had been an option within that tradition. What was new at this point was discussion of a specific velocity of transmission, which could in turn be related to the Weber law: If one assumed that the propagation occurred with a velocity specified by the ratio of units, c, and if one took into account how propagation at that velocity would affect the forces between moving charges, one could arrive, by a process of approximation, at something like the Weber force law. In that way, what appeared in Weber's law as a velocity-dependent force was explained as the result of a propagated force that did not have an intrinsic and primitive dependence on velocity, but gained its velocity dependence through effects contingent on the time of propagation. This promised to avoid the energetic problems, while at the same time providing a physical interpretation of the appearance of a constant with dimensions of velocity in the Weber law. Such an analysis on the basis of a retarded potential had been undertaken, in particular, by Carl Neumann in 1868, and Neumann's paper was cited by Helmholtz.[34] The notion of the retarded potential in fact had in it the seeds of an approach to the electromagnetic theory of light, and a theory developed in that way, independent of Maxwellian considerations, was proposed by Ludwig Lorenz, of the University of Copenhagen, in 1866. Helmholtz, however, did not adopt that approach, and the notion of propagation of electrical effects at the speed of light thus furnished for Helmholtz suggestive background for his appreciation of Maxwell's approach, rather than an independent route to the electromagnetic theory of light.[35]

Helmholtz's theory was, in its foundations, squarely within the charge-interaction tradition: he assumed the existence of primitive, microscopic electric charges; he took macroscopic charge to consist in the local accumulation of these microscopic charge carriers; he viewed electric current as the convection of these charge carriers; and he took electromagnetic interactions to be instantaneous actions at a distance – thus, in the foundations at least, adopting the most extreme form of the action-at-a-distance point of view. The crucial step in adapting that foundation to accommodate Maxwellian elements was to assume that both matter and ether were polarizable media, containing mobile charges that gave rise to electric dipoles under the influence of electromotive force. Calculations similar to Kirchhoff's of 1857 showed that starting with instantaneous action at a distance, in a medium containing mobile charge carriers, one would get waves of current (conduction current in Kirchhoff's case, and polarization current in Helmholtz's case) propagating, given the right parameters, with a velocity comparable to the velocity of light; associated with Helmholtz's waves of polarization current in the ether, in particular, would be electrical and magnetic forces acting on the currents, corresponding closely to the electric and magnetic fields in Maxwell's electromagnetic waves. The difficult point, in Helmholtz's theory, was that the velocity of these ether waves would equal the ratio of units only in the limit of infinite polarizability of the ether. If one considers again the paradigmatic case of the charging capacitor circuit, with only free ether between the plates, one sees that the displacement current can be regarded as a polarization current, and in the limit of infinite polarizability, the mathematics of Ampère's law will work out correctly – the total current will be solenoidal. In that situation, however, the Poisson bound charges on the surfaces of the ether next to the plates will exactly cancel the true charges on the plates, leaving no apparent source for the electrical effects in the space between the plates. The awkwardness of Helmholtz's formulation was perceived contemporaneously; indeed, Helmholtz's rendition of the theory gave an account of the relationship between electric charge and electric current that suffered from difficulties quite parallel to those encountered in Maxwell's own version of the theory. For the Continental audience, however, the Helmholtz reinterpretation, with its awkwardnesses, was much more comprehensible than the original Maxwell theory, and the Helmholtz formalism was the medium through which a generation of Continental researchers (from the 1870s into the 1890s) approached the content of Maxwell's theory.[36]

Aside from its coming across as more familiar and palatable to the Continental audience, Helmholtz's formalism – and especially later developments of it by H. A. Lorentz – turned out to have definite heuristic advantages. First, by viewing dielectric media as polarizable media in the traditional Poisson sense – that is, as made up of molecular components containing primitive charge carriers with mobility within the molecule –

Helmholtz's approach allowed for a treatment of the interaction between light and matter that was much more successful than that of the pure Maxwellian approach, yielding important new perspectives on the microscopic foundations of dispersion and of magnetooptical phenomena such as the Faraday rotation. Second, in the Helmholtz formalism, charges and currents functioned as the sources of electric and magnetic fields, rather than vice versa as in the Maxwellian perspective. In the Helmholtz formalism, it seems clear that by manipulating charges and currents, one would be able in turn to manipulate the electromagnetic field, and that would have furnished a basis for the hope that by manipulating laboratory electrical apparatus, one might be able to produce electromagnetic waves. In the Maxwellian perspective, on the other hand, it is harder to see exactly how it is possible to intervene in field processes, and the possibility of producing electromagnetic waves by artificial means would not have been quite so evident. That may be why it was Heinrich Hertz, a student of Helmholtz, who first was successful at generating electromagnetic waves with electrical apparatus, rather than Maxwell or a Maxwellian.[37]

4   *Toward the classical synthesis*

Heinrich Hertz, in 1888, was able to generate and detect electromagnetic waves with laboratory electrical apparatus, and that was received as final verification of the basic soundness of the concept of the displacement current and the formalism of transverse electromagnetic waves based on it. Hertz's discovery also gave strong impetus to the program of fashioning an electromagnetic theory that would be more faithful to the Faraday-Maxwell view of the field as an independent entity than was Helmholtz's formalism, while preserving the emphasis on primitive electrical charge carriers that was the heuristic strength of Helmholtz's approach. That led, in the years around 1900, to a final synthesis of the field-primacy and charge-interaction viewpoints, eventuating in the formulation of classical electromagnetic theory in its enduring form. Full consideration of the complexities of this culminating period in the development of classical electromagnetic theory, intertwined as it is with the history of the special theory of relativity, would be beyond the scope of this book; my purpose here will be to indicate, briefly and synoptically, some of the main themes in these developments.[38]

Helmholtz's laboratory in Berlin had become, by the 1880s, a center for experimentation relevant to the new electromagnetic theory and the new optics based on it. Henry Rowland and Albert Michelson journeyed from the United States to embark on the crucial phases of their research programs there, and Heinrich Hertz there began his work that led to the production of electromagnetic waves. Helmholtz, in 1879, had suggested to the Berlin Academy of Science that they offer a prize for a successful experimental test of the auxiliary assumptions that had to be added to the Continental charge-interaction electrodynamics in order to get the substance of Maxwell's theory. Those assumptions, according to Helmholtz, were as follows:

> First, that changes of dielectric polarisation in non-conductors produce the same electromagnetic forces as do the currents which are equivalent to them; secondly, that electromagnetic forces as well as electrostatic are able to produce dielectric polarisations; thirdly, that in all these respects air and empty space behave like all other dielectrics.

Helmholtz then "drew [Hertz's] attention to this problem" – the problem was no doubt intended for Hertz all along – "and promised that [he] should have the assistance of the Institute in case [he] decided to take up the work."[39]

Take up the work Hertz did, but the problem proved recalcitrant, and it was only eight years later (in 1887–8), after Hertz had moved on to Karlsruhe, that his efforts were to meet with success. Using tuned oscillating circuits, Hertz first was able to show – through the effects that dielectric materials had on those circuits – that polarization currents in dielectric materials produced magnetic effects; he had in that way verified the first of Helmholtz's list of assumptions. At that point, Hertz decided that it would be most efficacious to bypass the rest of Helmholtz's list and jump ahead to the main question, namely, verification of the propagation of macroscopically generated electromagnetic waves in air or vacuum. Hertz was able to adapt his tuned circuits to the task, first detecting in that way interference phenomena between waves in air and waves on a wire, and then going on to detect interferences between a wave proceeding directly from the oscillator and its reflection from a metal sheet.[40]

The generation and detection, in that manner, of electromagnetic waves was widely accepted as an *experimentum crucis,* justifying experimentally both Maxwell's introduction of the displacement current into the electromagnetic field equations and his founding of an electromagnetic theory of light on those equations. The significance of Hertz's experiment was, of course, much broader than that, opening new vistas in both science and technology; work on the technological exploitation of electromagnetic waves was begun in the 1890s by Guglielmo Marconi and others. As concerned electromagnetic theory itself, the effect of Hertz's work, especially for the Continental community, was to encourage a return to a purer Maxwellianism – as opposed to Helmholtz's rendition – while at the same time not giving up the power of Continental views in dealing with the interaction between light and matter. In the work of Lorentz and others, this led to an electromagnetic theory in which primitive, microscopic charge carriers resident in material bodies were regarded as the sources of electromagnetic effects – that was the legacy of the charge-interaction tradition – but those effects were to be regarded as propagated through a true Maxwellian field: Helmholtz's picture of an etherial medium populated with electric charges was relinquished, and electromagnetic waves in vacuum were now to be regarded not as waves of polarization current associated with these ubiquitous charges but rather as alternating electric and magnetic fields in the Maxwellian sense. What resulted was a dualistic electromagnetic theory in which both charges and fields were regarded as primitive and independent elements. Concurrently, in British work – especially the work of Joseph Larmor – primitive electric charges were appended to the basically Maxwellian account of fields, approaching by a different route the same dualistic theory. The result of this convergence was, finally, modern classical field theory, in which both electromagnetic fields and electric charges were to be regarded as primitive and fundamental, and elements of both the field-primacy and the charge-interaction traditions thus merged into a single theoretical approach.[41]

The dualistic nature of the resulting theory can be illustrated by reference, once more, to the charging capacitor circuit. The electric current in the wire is to be regarded as the translatory motion of primitive, microscopic electric charges in the conducting medium, and the charges on the plates as the accumulations or deficiencies of these primitive charge carriers. The displacement current in the space between the plates is understood as a field process, in which the changing electric field – driven by the changing charges on the plates that are its sources – constitutes the source of a magnetic field, without itself being reducible to any motion of electric charge.[42]

Conclusion

In the formulation of classical electromagnetic theory that emerged around the turn of the twentieth century, the displacement current and the electromagnetic theory of light had been completely emancipated from their original context in the theory of molecular vortices, and the field-primacy approach had been integrated with elements of the charge-interaction approach to yield a dualistic theory in which both charged particles and electromagnetic fields were seen as fundamental elements. As for views of the nature of the field itself, opinion around 1900 still located it in some kind of an ether, variously regarded as mechanical, electromagnetic, or sui generis. Less and less was said, however, about the particulars of the nature of that ether. Maxwell himself had established the trend of treating the ether in a more agnostic and abstract manner, in the later reformulations of his electromagnetic theory. In addition, the combination of field-primacy and charge-interaction elements in the new synthesis led to conceptual difficulties, which called into question the hope of arriving at a coherent and definitive physical interpretation of Maxwell's equations in term of etherial processes; this reinforced the trend established by Maxwell's own retreat from concrete representations of the ether. The classic expression of the *terminus ad quem* of that trend was Hertz's statement of 1892: "To the question, 'What is Maxwell's theory?' I know of no shorter or more definite answer than the following: – Maxwell's theory is Maxwell's system of equations." Ultimately, then, Maxwell's electromagnetic theory, which had been so thoroughly mechanical in its origins, came to play a central role in the decline of the mechanical worldview and the transition to a twentieth-century emphasis on the "system of equations" as the essence of physical theory.[43]

CONCLUSION

Three perennial issues in Maxwell scholarship have woven their way through our study of the origins of the displacement current and the electromagnetic theory of light in the context of the molecular-vortex model: The first concerns the relationship between Maxwell's accomplishments and the mechanical worldview, the second addresses the role of the field-primacy approach in the genesis of Maxwell's innovations, and the third concerns the unity and coherence of Maxwell's mechanical models and mathematical formalisms.

1   *Maxwell and the mechanical worldview*
We have seen that Maxwell's stance with respect to mechanical models and his use of them was conditioned by the confluence, in his educational background and scientific training, of Scottish (Edinburgh) and Cambridge traditions, with the former inclining toward an analogical interpretation of mechanical representations, and the latter toward a more ontologically committed approach, in which mechanical hypotheses were viewed as candidates for reality, and evidence of a hypothetico-deductive character was accepted as providing support for their realistic status. We have examined the movement of Maxwell's own ideas and practices concerning mechanical modeling: from an initial reliance on mechanical models as heuristic physical analogies – in the Scottish, skeptical vein – thence to the installation of the molecular-vortex model as the basis for a realistically intended physical theory – as countenanced in Cambridge methodology – and finally to an attenuated mechanism – representing Maxwell's own, carefully balanced position – in which the physical universe continued to be viewed as ultimately mechanical, but the possibility of coming to know the details of the mechanism receded indefinitely. We have seen that the molecular-vortex model played a crucial role in the genesis of the displacement current and the electromagnetic theory

of light; those innovations were, however, later completely divorced from the mechanical context in which they originated. The relationship between Maxwell's thought and the mechanical worldview was thus close and most significant for his work in electromagnetic theory, but that relationship was also complex, nuanced, and changing, and the Maxwellian legacy in electromagnetic theory was in fact ultimately to be enlisted against the mechanical worldview.[1]

Notwithstanding the nuances and changes over time of Maxwell's mechanical commitments, it is clear enough that at the time when he was developing the molecular-vortex model and drawing out of it the initial formulations of the displacement current and the electromagnetic theory of light, it was ontological concern and commitment in the tradition of Herschel and Whewell, rather than some instrumentalist agnosticism, that was driving his research program. Our analysis has shown, further, that the displacement current and the electromagnetic theory of light were not independent constructs loosely grafted to the molecular-vortex model, but rather organic parts of that model, growing naturally out of it: The basic mechanical parameters of the magnetoelectric medium were fixed in such a manner as to account for the relative strengths of electrical and magnetic forces, and the propagation of waves at the velocity of light followed directly from that; the new term in Ampère's law was introduced in order to allow for the accumulation of electric charge in the model, and the formulation of a complete and consistent set of electromagnetic equations, incorporating the displacement current, followed directly from that. A further measure of the strength of the connection between these two novelties and the molecular-vortex model is provided by the history of the difficult and fundamental reformulations that were required in order to finally break that connection. Thus, the linkage of the mechanical, molecular-vortex model with the displacement current and the electromagnetic theory of light is evident not only in the mechanical origins of those innovations but also in the difficulties associated with their subsequent emancipation from the mechanical matrix in which they originated.[2]

2   *Maxwell and the primacy of the field*

Maxwell's views on the relationship between charge and field were squarely in the field-primacy tradition, as established by Michael Faraday, in which the field was to be regarded as ontologically primary, with electric charge dependent on and emergent from the field, rather than vice versa; that view is contrary to the assumptions of modern classical field theory and therefore has presented interpretive problems for the modern student of Maxwell's work. The treatment of the relationship between charge and field in the molecular-vortex model is illuminating in this connection because it furnishes a concrete articulation of the field-primacy approach: In the model, charge – in the form of the idle-wheel particles – is in fact primitively a part of the field and in the field, but it is normally uniformly distributed throughout the field, and in that state manifests no effects. Effects are manifested when the medium undergoes elastic distortion in such a way as to displace and concentrate the charge associated with the field; the elastic deformations (or actually the restoring forces associated with them) then constitute the electric field, and the associated accumulations of idle-wheel particles constitute macroscopic electric charge. In that account of the relationship between field and charge, charge is emergent from the field at the macroscopic level, in consonance with the field-primacy orientation, but there are primitive charge carriers at the microscopic level, and in that the model falls short of complete faithfulness to Faraday's views. Maxwell's final formulation of the primacy of the field, in the *Treatise on Electricity and Magnetism*, was more faithful to the original Faraday perspective. In the *Treatise* there were no primitive, microscopic charge carriers, and charge was seen as emergent from a pattern of displacement – a kind of stress or constraint – in the medium; discontinuities or other divergence points in the displacement field were manifested as electric charge, and the effect was not reified in terms of some supposed particles of electricity.

Maxwell's successive articulations of the relationship between field and charge formed the basis for his successive conceptualizations of the displacement current. In the molecular-vortex model, the displacement current and the changing electric field associated with it were oppositely directed, with the displacement current transporting the microscopic charge carriers so as to give rise to macroscopic electric charge. In the later conceptualization of the *Treatise*, the displacement current and the changing electric field associated with it were in the same direction, with the displacement current giving rise to the stresses in the medium whose divergence points were manifested as electric charge. Neither of these conceptualizations of the displacement current and the relationship between field and charge proved to be fully serviceable in the closing decades of the nineteenth century, and the transition to modern classical field theory around the turn of the twentieth century involved the compromise of extreme field-primacy principles. Protracted adherence to those principles on the part of the British Maxwellians served to isolate that community, to the point that Oliver Lodge, alluding to these issues in

1902, would call attention to the bond of "sympathetic understanding running through" the community of "British Physicists," as contrasted with their isolation from the "work of Physicists outside." The story of the British field-primacy approach in electricity and magnetism, from Faraday through Lodge and his fellow Maxwellians, is testimony to the strength and tenacity of the programmatic commitments of national scientific communities.[3]

3   *The unity and coherence of Maxwell's electromagnetic theory*

Maxwell's goal, from the outset, was to develop a complete and coherent account of electromagnetic phenomena. Cambridge methodology, as most forcefully articulated by John Herschel and William Whewell, emphasized comprehensiveness of coverage and unification of diverse phenomena as goals for – and criteria for the evaluation of – deep theory, whose purpose was to achieve an understanding of the true mechanism of nature. Moreover, the task of electromagnetic theory at the time, as generally perceived and as dramatically exemplified in the work of Wilhelm Weber, was to develop a unified account of the various classes of electromagnetic phenomena; Maxwell's mission as an advocate of the field-primacy approach required that he match or better Weber's impressive unification on the basis of the charge-interaction approach. Finally, whereas Scottish skepticism did not recommend precipitate haste toward grand unification, it did recommend that if the task was to be undertaken, it should be undertaken carefully and soberly. Thus, when Maxwell decided that the time had come for serious theorizing, Cambridge methodological criteria, Scottish sobriety, and the demands of the field-primacy mission all converged in requiring that the theory – that is, the theory of molecular vortices – give a coherent and unified account of the diversity of electromagnetic phenomena.

In order to ensure that the coherence of the theory of molecular vortices would be maintained as it was extended to embrace the full variety of electromagnetic phenomena, Maxwell, at each step, established linkages of mechanical variables that mirrored the linkages of electromagnetic variables specified by the electromagnetic equations; that, in turn, provided for the establishment of a consistent set of correspondences between electromagnetic and mechanical variables. It was in order to mend a break in the chain of linkages that Maxwell first undertook to modify Ampère's law: The integrity of the resulting chain of parallel linkages of electromagnetic and mechanical variables signaled the achievement of a complete and consistent theory at both the electromagnetic and mechanical levels; furthermore, by successfully linking electrical and magnetic forces, it allowed for the incorporation of the ratio of electrical units into the theory, thus providing the basis for the initial articulation of the electromagnetic theory of light. Thus, the coherent and unified coverage of electromagnetic phenomena for which Maxwell strove and which he achieved in the molecular-vortex model was a sine qua non for his signal innovations: the displacement current and the electromagnetic theory of light.

It was in the years after 1862, when Maxwell was attempting to recast his theory so as to free it from its original mechanical matrix, that he encountered serious problems leading to the breakdown of the coherence of the theory; these problems were manifested mathematically as inconsistencies in algebraic signs and were connected physically with the problem of the nature of electric charge. The problems were resolved, and the theory regained consistency – although still not its modern form – in the *Treatise on Electricity and Magnetism.* After Maxwell, there was another period of rethinking and reorientation, issuing finally in the synthesis of British and Continental traditions that constituted the ultimate redaction of classical electromagnetic field theory. Viewed in broad sweep, therefore, the development of classical field theory, in Maxwell's own work and thereafter, involved the formulation of a series of differing versions of the theory, each internally coherent; intervening between successive versions were periods of searching, reorientation, and some confusion. Dominating the whole process, however, was the search for a comprehensive and consistent electromagnetic theory; the displacement current – keystone of a consistent set of field equations – and the electromagnetic theory of light emerged out of that search.

4   *Reflections on intellectual history, confusion, and clarity*

By understanding Maxwell's initial conceptualizations and subsequent reformulations of the displacement current and the electromagnetic theory of light, as developed in the context of his nineteenth-century commitments to the mechanical worldview and the field-primacy approach, we have been able to regain a set of intermediate stages in the development of electromagnetic theory whose existence and integrity has not hitherto been properly appreciated. One of the most important contributions to our understanding of the past that can be made by the intellectual historian – and by the historian of science qua intellectual historian – is to set before us such intermediate stages; one of the most straightforward answers that can be given to the question of how our intellectual forebears got from one point to another in the development of their thinking is to exhibit an intermediate step or steps through which, and by means of which, the transition from the one point to the other was facilitated and accomplished.[4] (The critic of intellectual history might object that the interpolation of additional points between the two given points does not solve the problem in principle: One is still left with the irreducible leap from point to point. The intellectual historian will point out, in defense, that as the number of points increases, the sizes of the leaps decrease, the path is more precisely delineated, and the historical account becomes more satisfying. One can perhaps never regain the continuum, but then – as some would argue – neither can the mathematicians.[5])

Also, by understanding well the extent to which Maxwell's formulations of the displacement current and the electromagnetic theory of light differ from our modern formulations, we have been able to make sense of Maxwell's formulations and appreciate their internal coherence. Those who would judge, by measuring it against the modern formalism, that Maxwell's theory of molecular vortices was incoherent, must conclude that some of the most consequential innovations in the history of physics were born out of confusion; such a conclusion would support the "sleep-walking" model of scientific innovation.[6] On the other hand, if one judges, by understanding it in context, that the theory was coherent, one can conclude that the displacement current and the electromagnetic theory of light were, to a great extent, if not completely, products of clear thinking and rational analysis. In other words, rather than seeing Maxwell as a muddleheaded twentieth-century physicist who happened to live in the nineteenth century, one is able to see Maxwell as a clearheaded nineteenth-century physicist whose accomplishments grew out of, and are comprehensible within, the nineteenth-century context.

APPENDIX 1

Draft of "On Physical Lines of Force," a fragment

Few working papers survive from the period when Maxwell was working on "Physical Lines." My own search of relevant archives, as well as the more exhaustive search conducted by Peter Harman in connection with his edition in progress of *The Scientific Letters and Papers of James Clerk Maxwell,* 3 vols. (Cambridge University Press, 1990–), turned up nothing beyond the material to which attention is directed by A. E. B. Owens's handlist to Add. MSS 7655 at the University Library, Cambridge. Of a set of five folios constituting Add. MSS 7655, V, c/8, two folios clearly correspond to the period when Maxwell was working on "Dynamical Theory" (see Appendix 2), and a third, dealing with "Helmholtz's Wirbelfäden [Vortex Filaments]," appears to date from a later period as well (1864–70 – Peter Harman, private communication; see also *Letters and Papers of Maxwell, 2*). The two remaining folios (both blank verso) evidently are associated with "Physical Lines." One of these clearly relates to the treatment of motional electromotive forces that appears in "Physical Lines," Part II, 476–85; this draft fragment refers explicitly to "equations (55)," which appear on p. 476 of the published version, and arrives at a form of the published equations (77), on p. 482.

The remaining folio is quite informative concerning various aspects of Maxwell's work on the molecular-vortex model; a photograph of it is presented in Fig. A1.1, and I here transcribe it in full (cf. also the transcription in Harman, ed., *Letters and Papers of Maxwell, 1*, 693). In this transcription, ⟨ ⟩ indicates material that was deleted in the original, with ⟨⟨ ⟩⟩ used for a large deletion, ⟨?⟩ indicating undecipherable deleted material, and ⟨ ⟩xxx, without a space, indicating material overwritten by xxx; [ ] indicates material that was inserted in the original; { } indicates my inserted numbering of paragraphs and equations. Maxwell's script is rendered in roman type, and his block capitals as boldface.

Prop XII

To find the angular momentum of a vortex.

The angular momentum of any system about an axis is the sum of the products of each particle multiplied ... by the area ... I describe about that axis in unit of time or of A = the angular momentum about the axis of x

$$A = \sum dm \left(y \frac{dz}{dt} - z \frac{dy}{dt}\right)$$

As we do not know the law of distribution of density in the vortex, we shall determine the relation between the angular momentum and the Energy of the vortex on a

[crossed out: different assumption]

Let the motion of the vortex be such that the angular velocity is the same throughout and let v be the linear velocity at the circumference, then if A is the angular momentum and E the energy

$$A = \sum dm \, r^2 \omega$$

$$E = \frac{1}{2} \sum dm \, r^2 \omega^2$$

Making ω constant we have $\quad A = \frac{2E}{\omega} \quad$ I.E. ?

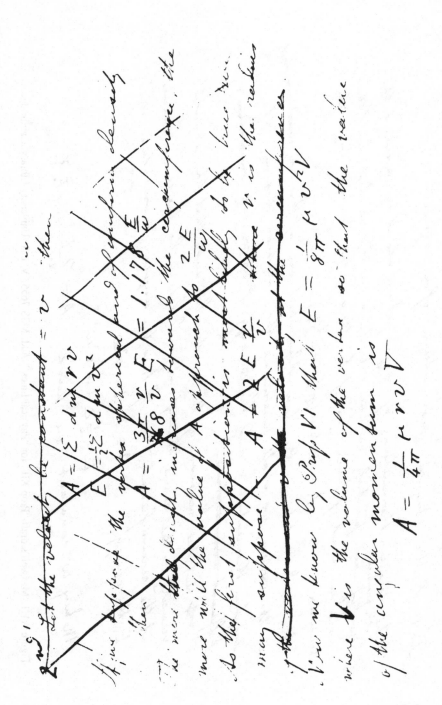

Figure A1.1 (continued).

Prop XII

To find the Angular momentum of a vortex.

{1} The Angular Momentum of any system about an axis is the sum [of the products] of each particle multiplied by the area it describes about that axis in unit of time or if **A** is the angular momentum about the axis of x

$$\mathbf{A} = \Sigma \, dm \left(y\frac{dz}{dt} - z\frac{dy}{dt} \right) \qquad \{A1.1\}$$

As we do not know the law of distribution of density ⟨and velocity⟩ in the vortex, we shall determine the relation between the Angular Momentum and the Energy of the vortex ⟨on several different assumptions⟩.

{2} ⟨1st⟩ Let the motion of the vortex be such that the angular velocity is the same throughout, and let the v be the linear velocity at the circumference, then if **A** is the angular momentum and **E** the energy

$$\mathbf{A} = \Sigma \, dm \, r^2\omega \qquad \{A1.2\}$$

$$\mathbf{E} = \tfrac{1}{2} \Sigma \, dm \, r^2\omega^2 \qquad \{A1.3\}$$

Making ω constant we have

$$\mathbf{A} = \frac{2\mathbf{E}}{\omega} = 2\mathbf{E}\,\frac{r}{v} \qquad \{A1.4\}$$

{3} ⟨⟨2nd Let the velocity be constant = v then

$$\mathbf{A} = \Sigma \, dm \, rv \qquad \{A1.5\}$$

$$\mathbf{E} = \tfrac{1}{2} \Sigma \, dm \, v^2 \qquad \{A1.6\}$$

If we suppose the vortex spherical and of uniform density

$$\text{then } \mathbf{A} = \frac{3\pi}{\langle 16 \rangle \, 8}\frac{r}{v}\,\mathbf{E} = 1.178\,\frac{\mathbf{E}}{\omega} \qquad \{A1.7\}$$

The more ⟨?⟩the density increases towards the circumference, the more

will the value of **A** approach to $\dfrac{2\mathbf{E}}{\omega}$

As the first supposition is most likely to be true we may suppose

$$\mathbf{A} = 2\mathbf{E}\,\frac{r}{v} \qquad \{A1.8\}$$

where r is the radius of the vortex and v the velocity at the circumference.⟩⟩

{4} Now we know by Prop VI that

$$\mathbf{E} = \frac{1}{8\pi} \mu v^2 \mathbf{V} \qquad \qquad \{A1.9\}$$

where ⟨v⟩V is the volume of the vortex so that the value of the angular momentum is

$$\mathbf{A} = \frac{1}{4\pi} \mu r v \mathbf{V} \qquad \qquad \{A1.10\}$$

This manuscript "Prop XII," subtitled "To find the Angular momentum of a vortex," parallels in basic content the published "PROP. XVIII. – To find the angular momentum of a vortex," in Part IV of "Physical Lines," pp. 508–9: Paragraph {1} of the manuscript (given Maxwell's insertions and deletions as indicated) corresponds closely to the published version, with equations {A1.1}, {A1.2}, and {A1.3} corresponding exactly; subsequent equations and the conclusion differ slightly in notation and arrangement, although not in content. The numbering of the manuscript proposition as XII, however, would locate it just after the last proposition in Part II of "Physical Lines," numbered "PROP. XI," on p. 481. The context in Part II of "Physical Lines" for such a calculation of angular momentum would have been furnished by the discussion, on pp. 485–6, of the diameters of the vortices and their angular momenta, in connection with Maxwell's report there of his effort to detect the angular momenta of the vortices experimentally (see Chapter 2, Section 3). If there had been any definite results to report, the manuscript "Prop XII" would have been required for their interpretation; because all that Maxwell had to report was that he had "not yet fully tried the apparatus," this "Prop XII" presumably was not needed and hence was not published. A need for this calculation arose again in Part IV, in connection with the treatment of the Faraday rotation, and Maxwell resurrected it there as Proposition XVIII. One of the main differences between the manuscript "Prop XII" and the published Proposition XVIII is notational: In the manuscript "Prop XII," the variables r, the radius of a vortex, and v, the circumferential velocity, were used in many of the equations; that was appropriate to a context in which the concern was with the concrete, mechanical aspects of the model (e.g., the sizes and angular momenta of the vortices). In the published Proposition XVIII, r and v were eliminated from some of the calculations, by making use of the angular momentum, $\omega = v/r$, and the x

component of the magnetic field strength, $\alpha = v$ (for rotations about the x axis); that was appropriate to Part IV of "Physical Lines," where the concern was more with electromagnetic variables and effects, and less with the specific mechanical characteristics of the substratum.

If the attribution of the manuscript "Prop XII" to the period when Maxwell was working on Part II of "Physical Lines" is correct, then it provides the most direct evidence (in paragraph {3} of the manuscript) that he was already thinking of the vortices as approximable to spheres at that time. This has bearing on the discussion of the shapes of the vortices in Chapter 3, Section 2, and helps to establish continuity between Part II of "Physical Lines" and Part III, where the vortices were explicitly pseudospherical.

The manuscript "Prop XII" is of interest also for the light it sheds on Maxwell's treatment of, and difficulties with, factors of 2: A factor of 16 is crossed out in favor of 8 in the denominator of {A1.7}; a factor of $\frac{1}{2}$ is squeezed in as an afterthought in {A1.6}; and "the area" in paragraph {1} becomes "twice the area" in the published Proposition XVIII. If this fragment is representative, misestimations and subsequent corrections of factors of 2 were common occurrences in Maxwell's work, and that would have provided fertile ground for the kinds of errors and adjustments discussed in Chapter 5, Section 4. Also relevant in this connection is a comparison of {A1.4} (or {A1.8}) with {A1.7}, which gives some feeling for the latitude in results allowed by different assumptions concerning the vortices – here a factor of $2/1.178 = 1.7$. Paragraph {3}, in which this occurs, also exhibits, in microcosm, a general trend in Maxwell's treatment of the molecular-vortex substratum: He began paragraph {3} by considering a second, alternative hypothesis concerning the specifics of the vortex rotation; he showed that the second hypothesis could be made to reduce to the first under certain circumstances; he observed that the first was "most likely" to be true anyway; and he ended by deleting any mention of the second hypothesis. This progressive streamlining of the assumptions of the model, converging on a definite and simple answer (2 was chosen as the coefficient to be carried forward, rather than 1.178 or an intermediate value), is reflective of the general process discussed in Chapter 5, Section 4; see also Chapter 3, Section 1, especially the discussion relating to equation (3.6). The choice of uniform density and angular velocity ({A1.4} or {A1.8} as against {A1.7}) in the manuscript "Prop XII" was then supported by the specification of a homogeneous elastic-solid material for the contents of the vortex cells in Part III of "Physical Lines," thus justifying the subsequent use of the proposition in this form in Part IV, as the published Proposition XVIII.

APPENDIX 2

Drafts of "A Dynamical Theory of the Electromagnetic Field"

Manuscript material relating to "Dynamical Theory" and of interest in connection with Chapter 6, Section 1, includes four pages in the Maxwell manuscript materials at the University Library, Cambridge (in Add. MSS 7655, V, c/8; c/11; and V, f/4), to which I shall make reference as follows: Three pages, apparently representing parts of an early draft of "Dynamical Theory," and corresponding to pp. 559–61, 568, and 569, I shall denote "[DT, A]," "[DT, B]," and "[DT, C]"; a fourth, apparently representing part of a later draft of "Dynamical Theory," and corresponding to p. 578, I shall denote "[DT, D]." The correspondences to the cited parts of "Dymanical Theory" are not in doubt; that A, B, and C are earlier is suggested by the numbering of the equations, which differs from the published version, whereas D agrees with the published version in numbering of equations. A and B are pages numbered 22 and 23 (evidently in Maxwell's hand), in V, c/8; C is a page numbered 24, in V, f/4, but helpfully identified in the handlist to Add. MSS 7655 as belonging with A and B. D is in V, c/11. Also of interest is the manuscript of "Dynamical Theory" that was submitted to the Royal Society and is preserved there – PT. 72.7 – to which I shall refer as "Dynamical Theory [MS]." (I am informed that much of this material will be published in Harman, ed., *Letters and Papers of Maxwell, 2.*)

Of particular visual interest is a part of "[DT, B]," a photograph of which is presented in Figure A2.1. This is a draft form of the published equations (E), "Dynamical Theory," 560, reproduced herein as equation (6.1). The negative sign intervening between *f, g, h* and *P, Q, R* in the draft disappears in the published version, and the sense of dilemma associated with this question of algebraic sign (see Chapter 6, Section 1) is conveyed visually in Maxwell's writing of the equations in the manuscript: The writing of the equals signs speaks tentativeness – they are wavering, made up of multiple penstrokes, and drawn out to three times

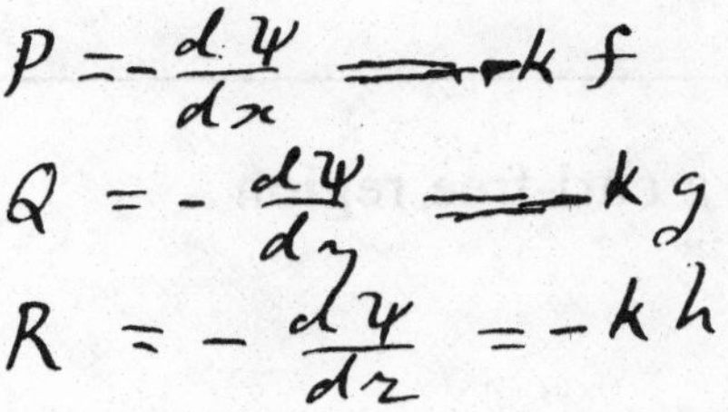

$$P = -\frac{d\Psi}{dx} \quad \text{=== } kf$$

$$Q = -\frac{d\Psi}{dy} \quad \text{=== } kg$$

$$R = -\frac{d\Psi}{dz} = -kh$$

Figure A2.1. Maxwell's draft "[DT, B]" for "Dynamical Theory," Add. MSS 7655, V, c/8, University Library, Cambridge.

normal length, as if in hesitation; the minus signs are also nervously worked over. By the time he got to the third component equation, Maxwell wrote with more resolve – only to reverse himself again in the published version.

APPENDIX 3

Vortex rotations in a curl-free region

Figures 4.8 and 4.9a depict schematically the vortex rotations and idle-wheel translations associated with a uniform current density inside of a long, straight wire with uniform circular cross section. Inside the wire, the magnetic field grows linearly with distance from the axis; because of this, neighboring vortices rotate with different angular velocities; this engenders motion of the idle-wheel particles interposed between the vortices, constituting a nonzero current density **J**; and the inhomogeneity of the magnetic field **H** is associated with a nonzero value for **curl H**, which is equal to the nonzero current density **J**.

Outside of the wire, the **H** field falls off as $1/r$, where r is the distance from the axis [E. R. Peck, *Electricity and Magnetism* (New York: McGraw-Hill, 1953), 214–17]. One might, at first thought, expect that because of this, neighboring vortices would rotate with different angular velocities, and this would engender motion of the idle-wheel particles, constituting a nonzero current density **J**. Even though the magnetic field **H** is inhomogeneous, however, **curl H** and hence **curl ω*** are zero outside the wire, and Maxwell's calculation leading to equation (3.7a) shows that in this situation there will be no net flux of the idle-wheel particles, and hence no current ι or **J**. The mathematics leading to this result is not in doubt, but how are we to understand it intuitively – how can there be no flux of idle-wheel particles in a situation in which the magnetic field is inhomogeneous and the angular velocities of neighboring vortices are therefore different?

In order to get some feeling for this, it is necessary to give attention to the disposition of the vortex rotations in the plane perpendicular to the axis of the wire, as in Figure A3.1. Here, let I denote (the axis of) the wire, and let **H, H, H** denote the magnetic field lines in the region outside the wire. Let the vortices be schematically considered as spherical, so that their sections in the plane will be circular; the vortices can be thought of

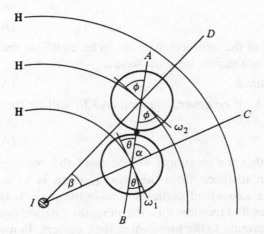

Figure A3.1. Vortex rotations in the curl-free region outside of a conducting wire.

as in hexagonal array, square array, or other – it does not matter to the current argument. Let the two smaller circles in the figure then represent an arbitrary pair of neighboring vortices in the array – coupled by a single idle-wheel particle, just as in Figure 4.8 or 4.9a – with the rotation axes, in the plane, of vortices 1 and 2 (tangent to the field lines **H, H**) denoted respectively by ω_1 and ω_2, the magnitudes of their angular velocities. Let the line AB, going through the centers of the vortices, be inclined at an arbitrary angle α to line IC from the axis of the wire I to the center of the closer vortex, vortex 1, where $0 \le |\alpha| \le \pi/2$, and let β be the angle between lines IC and ID through the centers of the two vortices. The axes ω_1 and ω_2 of the vortices are perpendicular respectively to IC and ID, so that the angle between ω_1 and ω_2 is also β, and the angles θ and ϕ that ω_1 and ω_2 respectively make with any given line, such as AB, are related by

$$\phi = \theta + \beta \tag{A3.1}$$

so that for the given ranges of the angles, and except for the limiting case $\beta = 0$,

$$\phi > \theta \tag{A3.2}$$

$$\sin \phi > \sin \theta \tag{A3.3}$$

The idle-wheel particle between the two vortices makes contact with points on their surfaces at angles θ and ϕ from their rotation axes, respectively, so that if the radii of the vortices are r, the magnitudes of the linear velocities, v_1 and v_2, of those points on the surfaces are given by

$$v_1 = r\omega_1 \sin\,\theta \tag{A3.4}$$

$$v_2 = r\omega_2 \sin\,\phi \tag{A3.5}$$

Thus, if the magnitudes of the two velocities are to be equal, so that the idle-wheel particle does not move, we must have

$$\omega_1 \sin\,\theta = \omega_2 \sin\,\phi \tag{A3.6}$$

Given the inequality (A3.3), however, equation (A3.6) will be satisfied only if

$$\omega_2 < \omega_1 \tag{A3.7}$$

What this tells us is that the rotational velocities of the vortices, ω, *must* be falling off with distance from the wire if there is to be no translatory motion of the idle-wheel particles outside the wire. It is the curvature of the field lines that requires this, inclining the rotation axes of neighboring vortices differently to the line joining their centers. In mathematical terms, any component of the **curl** operator involves two spatial derivatives, each of which may be nonzero, while they cancel to give a zero **curl** component; in this way, the curvature of the field lines compensates for the falloff in intensity of the field in order to yield a zero **curl.** (Taking this into account in the analyses associated with Figures 4.8 and 4.9a would not change the results obtained there for the region within the wire. First, the mathematics guarantees this. Second, in intuitive terms, a further analysis in the plane perpendicular to the symmetry axis of the wire, such as undertaken in this appendix, would introduce two compensating corrections: The rotational velocities of the vortices would increase less rapidly from one vortex to the next in going outward from the axis, but the provision in this analysis for a range of angles $\beta \neq 0$ – as in Figure A3.1 – would serve to bring the vortices themselves closer in than is allowed for by the analysis of Chapter 4 – where $\beta = 0$ throughout – and this would cancel the first effect, thus regaining the linear increase within the wire.)

NOTES

Introduction

1 James Clerk Maxwell, "On Physical Lines of Force, Part I: The Theory of Molecular Vortices Applied to Magnetic Phenomena," *Philosophical Magazine, 21* (No. 139, March 1861), 161–75; "On Physical Lines of Force, Part II: The Theory of Molecular Vortices Applied to Electric Currents," *Philosophical Magazine, 21* (No. 140, April 1861), 281–91, and *21* (No. 141, May 1861), 338–45; "On Physical Lines of Force, Part III: The Theory of Molecular Vortices Applied to Statical Electricity," *Philosophical Magazine, 23* (No. 151, January 1862), 12–24; "On Physical Lines of Force, Part IV: The Theory of Molecular Vortices Applied to the Action of Magnetism on Polarized Light," *Philosophical Magazine, 23* (No. 152, February 1862), 85–95. Reprinted under the common title "On Physical Lines of Force," in W. D. Niven, ed., *The Scientific Papers of James Clerk Maxwell, 2* vols. (1890; New York: Dover, 1965 [2 vols. in 1]), *1*, 451–513; quoted material on p. 489.

2 For a recent synoptic review of the Maxwell literature, emphasizing its problems and its nonprogressive nature, see Norton Wise, "The Maxwell Literature and British Dynamical Theory," *Historical Studies in the Physical and Biological Sciences, 13* (1982), 175–205. Seminal for me in posing the questions was Joan Bromberg, "Maxwell's Displacement Current and His Theory of Light," *Archive for History of Exact Sciences, 4* (1967), 218–34, and idem, "Maxwell's Electrostatics," *American Journal of Physics, 36* (1968), 142–51.

3 Pioneering for the modern study of these questions are Joseph Turner, "Maxwell on the Method of Physical Analogy," *British Journal for the Philosophy of Science, 6* (1955), and idem, "Maxwell on the Logic of Dynamical Explanation," *Philosophy of Science, 23* (1956), 36–47; see also Robert Kargon, "Model and Analogy in Victorian Science, Maxwell's Critique of the French Physicists," *Journal of the History of Ideas, 30* (1969), 423–36; P. M. Heimann [Harman], "Maxwell and the Modes of Consistent Representation," *Archive for History of Exact Sciences, 6* (1970), 171–213; Mary Hesse, "Logic of Discovery in Maxwell's Electromagnetic Theory," in R. N. Giere and Richard S. Westfall, eds., *Foundations of Scientific Method: The Nineteenth Century* (Bloomington: Indiana University Press, 1973), 86–114; and A. F. Chalmers, "Maxwell's Methodology and His Application of It to Electromagnetism," *Studies in History and Philosophy of Science, 4* (1973), 107–64. My thinking about models and analogies has been much facilitated by my

reading in Peter Achinstein, *Concepts of Science: A Philosophical Analysis* (Baltimore: Johns Hopkins University Press, 1968), esp. 203–25. The *locus classicus* on the inutility of Maxwell's models is Pierre Duhem, *The Aim and Structure of Physical Theory* (1954; New York: Atheneum, 1962), 55–104, esp. 98.

4   See, e.g., Bromberg, "Maxwell's Electrostatics"; Hesse, "Logic of Discovery"; Chalmers, "Maxwell's Methodology"; Harman, "Consistent Representation."

5   The *locus classicus* arguing the incoherence of Maxwell's theories is Pierre Duhem, *Les Théories Electriques de J. C. Maxwell: Etude Historique et Critique* (Paris: A. Hermann, 1902); cf., e.g., Bromberg, "Displacement Current," and Wise, "British Dynamical Theory." Classical on the other side is Ludwig Boltzmann's commentary on Maxwell's paper "On Physical Lines of Force," in Ludwig Boltzmann, ed. and tr., *Ueber physikalische Kraftlinien,* Ostwald's Klassiker der exacten Wissenschaften, No. 102 (Leipzig: Wilhelm Engelmann, 1898).

6   This conclusion is adumbrated, e.g., in Daniel M. Siegel, "Completeness as a Goal in Maxwell's Electromagnetic Theory," *Isis, 66* (1975), 361–8, and, in parallel, in Ole Knudsen, "The Faraday Effect and Physical Theory, 1845–1873," *Archive for History of Exact Sciences, 15* (1976), 235–81.

7   The analysis leading to this conclusion begins from Boltzmann, ed., *Ueber physikalische Kraftlinien,* and has points of contact with, e.g., Bromberg, "Maxwell's Electrostatics."

8   For a classic exposition of the dangers attendant upon the translation of historical mathematical arguments into modern notation, see Sabetai Unguru, "On the Need to Rewrite the History of Greek Mathematics," *Archive for History of Exact Sciences, 15* (1975–6), 67–114, where, in particular, the translation of ancient Greek geometrical discourse into modern algebraic symbolism is decried. It seems to me, however, that any translation, rendering, paraphrase, summary, or like manipulation of a historical text is subject to the same danger of "imposing on that [text] a content it does not in fact possess" (in the words of Michael S. Mahoney, as quoted by Unguru, p. 67). Unguru himself does paraphrase, showing an awareness of the possible dangers of that exercise when he asserts that "my paraphrase is *faithful* to Euclid's way of reasoning" (p. 98, emphasis Unguru's). Indeed, the application of any tools of representation or analysis to historical texts is attended by the danger of distortion; we would, however, no more give up the use of all intellectual tools in historical discourse than we would give up the use of all knives or hammers because of the risk of bodily injury. What we are left with, then, is the need to judge the balance of risk and advantage in each use of each tool; in particular, the judgment that the translation of Greek geometry into algebraic symbolism has been more harmful than beneficial to the historical study of ancient mathematics cannot support any general a priori judgment concerning the appropriateness in historical discourse of the translation of mathematical arguments into alternative notation. As for the concern that such translation will result in "Whig history," the reader will have to judge a posteriori to what extent I have been successful in "look[ing] at the past through sympathetic and understanding eyes and . . . achiev[ing] a reconstruction which does no patent violence to that which is to be reconstructed" (p. 68).

1. The background to Maxwell's electromagnetic theory

1   John L. Heilbron, "The Electrical Field before Faraday," in G. N. Cantor and M. J. S. Hodge, eds., *Conceptions of Ether: Studies in the History of Ether Theories,*

1740–1900 (Cambridge University Press, 1981), 187–213, esp. 196–9. See also Robert C. Stauffer, "Speculation and Experiment in the Background of Oersted's Discovery of Electromagnetism," *Isis, 48* (1957), 33–50; Thomas S. Kuhn, "Energy Conservation as an Example of Simultaneous Discovery," in idem, *The Essential Tension* (University of Chicago Press, 1977), 66–104; L. Pearce Williams, *The Origins of Field Theory* (New York: Random House, 1966), esp. 32–63.

2   R. A. R. Tricker, *Early Electrodynamics: The First Law of Circulation* (Oxford: Pergamon Press, 1965), esp. 140–54 (Ampère's early papers) and 181 (the force law). Heilbron, "Electrical Field," 199–203; L. Pearce Williams, *Michael Faraday: A Biography* (New York: Basic Books, 1964), 142–8. Edmund Whittaker, *A History of the Theories of Aether and Electricity*, rev. and enl. ed., 2 vols. (1951; New York: Humanities Press, 1973), *1*, 57–65.

3   James Clerk Maxwell, *A Treatise on Electricity and Magnetism*, 2 vols. (Oxford: Clarendon Press, 1873); 3rd ed., 2 vols. (1891; New York: Dover, 1954), *2*, 162 (1st ed.), 175 (3rd ed.). Isaac Newton, *Opticks*, 4th ed. (1730; New York: Dover, 1952), 350–3 (Queries 21, 22); cf. Richard S. Westfall, *Never at Rest: A Biography of Isaac Newton* (Cambridge University Press, 1980), 271, 305–8, 373–7, 464–5, 505–9, 644, 793–4. Heilbron, "Electrical Field," 203–4; Williams, *Faraday*, 148–50. Robert Fox, "The Rise and Fall of Laplacian Physics," *Historical Studies in the Physical and Biological Sciences, 6* (1974), 89–136, esp. 116–17, defines a "Laplacian" tradition in French physics that closely approximates the French background for Ampère described herein, and he discusses Ampère's falling away from that tradition; Fox's categories and my own are, however, slightly oblique to each other, which leads to different emphasis in placing Ampère. On the intricacies of the "action at a distance" question, see, e.g., A. E. Woodruff, "Action at a Distance in Nineteenth-Century Electrodynamics," *Isis, 53* (1962), 439–59, esp. 440–2, and M. Norton Wise, "German Concepts of Force, Energy, and the Electromagnetic Ether: 1845–1880," in Cantor and Hodge, eds., *Conceptions of Ether*, 269–307.

4   P. M. Harman, *Energy, Force, and Matter: The Conceptual Development of Nineteenth-Century Physics* (Cambridge University Press, 1982), 19–21. Williams, *Faraday*, 11–20, esp. 13, 66–9, 151–68; quotation of Faraday to Ampère, 2 February 1821, on p. 168. Cf. L. Pearce Williams, "Faraday and Ampère: A Critical Dialogue," in David Gooding and Frank A. J. L. James, eds., *Faraday Rediscovered: Essays on the Life and Work of Michael Faraday, 1791–1867* (New York: Stockton Press, 1985), 83–104.

5   Michael Faraday, "On Some New Electro-Magnetical Motions, and on the Theory of Magnetism" (1821), in *Experimental Researches in Electricity*, 3 vols. (1839–55; New York: Dover, 1965 [3 vols. in 2]), *2*, 127–47, esp. 132: "It is, indeed, an ascertained fact, that the connecting wire has different powers at its opposite sides; or rather, each power continues all round the wire, the direction being the same, and hence it is evident that the attractions and repulsions of M. Ampère's wires are not simple, but complicated results."

6   Faraday, "Electro-Magnetical Motions," 143 – "filings sprinkled on paper"; idem, *Experimental Researches, 1*, 32n (Series I, 1831) – "lines of magnetic forces"; *3*, 81 (Series XXI, 1845) – "magnetic field"; *3*, 397–401 (Series XXIX, 1851), esp. 397 – "§37. Delineation of Lines of Magnetic Force by Iron Filings," and diagrams, Plate III.

7   Faraday, *Experimental Researches, 1*, 360–416 (Series XI, 1837), esp. 360–7, 409–13; *1*, 515–32 (Series XIII, 1838), on 515, 523.

8   For a review of Faraday's field theory and the literature on it, see Harman, *Energy, Force, Matter*, 72–9, 166–7; Harman discusses the primacy of the field for Faraday in connection with matter theory. A *locus classicus* on the various views of the relationship of charge and field is Heinrich Hertz, *Electric Waves*, tr. D. E. Jones (1893; New York: Dover, 1962), 20–8.

9   See, e.g., Wise, "German Concepts"; Woodruff, "Action at a Distance"; Daniel M. Siegel, "Thomson, Maxwell, and the Universal Ether in Victorian Physics," in Cantor and Hodge, eds., *Conceptions of Ether*, 239–68, esp. 239–40, 261–3; Jed Z. Buchwald, *Maxwell to Microphysics: Aspects of Electromagnetic Theory in the Last Quarter of the Nineteenth Century* (University of Chicago Press, 1985), passim, and review thereof by Daniel M. Siegel in *Science, 232* (1986), 532–3.

10 Faraday, *Experimental Researches, 1*, 1–41 (Series I, 1831). Williams, *Faraday*, 137–211, 283–99, 381–407. Heilbron, "Electrical Field," 204–7.

11 James Clerk Maxwell, "On Faraday's Lines of Force" (read 10 December 1855 and 11 February 1856), *Transactions of the Cambridge Philosophical Society, 10* (pub. 1864), 27–65, reprinted in Niven, ed., *Scientific Papers of Maxwell*, 155–229, on 207.

12 Wise, "German Concepts," 276–83; Woodruff, "Action at a Distance," 445–6; W. Thomas Archibald, "'Eine sinnreiche Hypothese': Aspects of Action-at-a-Distance Electromagnetic Theory, 1820–1880," dissertation, University of Toronto, 1987, 140–261 (Ch. 3, pagination may vary slightly); L. Rosenfeld, "The Velocity of Light and the Evolution of Electrodynamics," *Nuovo Cimento, 4 (Suppl.)* (1957), 1633–4 (my identification of Weber's constant c as c_w, in order to distinguish it from the modern constant c, follows Rosenfeld – see further Chapter 5, Section 1).

13 Faraday, *Experimental Researches, 1*, 16–22 (Series I, 1831), on 16.

14 Faraday, *Experimental Researches, 1*, 32–3 (Series I, 1831); *3*, 328–70 (Series XXVIII, 1851), esp. 346.

15 Faraday, *Experimental Researches, 1*, 360–416 (Series XI, 1837); *3*, 1–160 (Series XX–XXIII, 1845–50); *3*, 169–273 (Series XXV–XXVI, 1850). Williams, *Faraday*, passim.

16 Maxwell to Thomson, 20 February 1854, in Joseph Larmor, ed., "The Origins of Clerk Maxwell's Electric Ideas, as Described in Familiar Letters to W. Thomson," *Proceedings of the Cambridge Philosophical Society, 32* (1936), 695–750 [also published as *The Origins of Clerk Maxwell's Electric Ideas, as Described in Familiar Letters to William Thomson* (Cambridge University Press, 1937)], 697–8, on 697; also in P. M. Harman, ed., *The Scientific Letters and Papers of James Clerk Maxwell*, 3 vols. (Cambridge University Press, 1990–), *1*, 237–8. As the preparation of *Letters and Papers of Maxwell* paralleled my own work, I have not been able to make fullest use of it; I thank Peter Harman for making parts of it available to me prior to publication.

17 C. W. F. Everitt, *James Clerk Maxwell: Physicist and Natural Philosopher* (New York: Scribner's, 1975), 49–53; idem, "Maxwell's Scientific Creativity," in *Springs of Scientific Creativity: Essays on Founders of Modern Science*, eds. Rutherford Aris, H. Ted Davis, and Roger H. Stuewer (Minneapolis: University

of Minnesota Press, 1983), 73–141, esp. 83–106. David B. Wilson, "The Educational Matrix: Physics Education at Early-Victorian Cambridge, Edinburgh, and Glasgow Universities," in P. M. Harman, ed., *Wranglers and Physicists: Studies on Cambridge Mathematical Physics in the Nineteenth Century* (Manchester University Press, 1985), 12–48, esp. 22. Lewis Campbell and William Garnett, *The Life of James Clerk Maxwell*, 2nd ed., with a new preface and appendix with letters by Robert H. Kargon, The Sources of Science, No. 85 (1882; New York: Johnson Reprint, 1969), 105–222, esp. 107–9, 144–6. Harman, ed., *Letters and Papers of Maxwell, 1,* 1–12.

18 George Elder Davie, *The Democratic Intellect: Scotland and Her Universities in the Nineteenth Century* (Edinburgh University Press, 1961), 103–200. P. M. Harman, "Edinburgh Philosophy and Cambridge Physics: The Natural Philosophy of James Clerk Maxwell," in Harman, ed., *Wranglers and Physicists,* 202–24, esp. 204–8. Campbell and Garnett, *Life,* 154, 185. Everitt, "Maxwell's Creativity," 104–5, emphasizes the "impressive array of mathematical techniques" that the tripos candidate acquired and calls attention to "the advantage that accrues to a working scientist . . . from sheer familiarity with technique."

19 Harvey W. Becher, "William Whewell and Cambridge Mathematics," *Historical Studies in the Physical and Biological Sciences, 11* (1980), 1–48, on 6. Thomas S. Kuhn, "Mathematical versus Experimental Traditions in the Development of Physical Science," in Kuhn, *Essential Tension,* 31–65. P. M. Harman, "Newton to Maxwell: The *Principia* and British Physics," *Notes and Records of the Royal Society of London, 42* (1988), 75–96. Concerning some of the subtleties of the question of Edinburgh versus Cambridge mathematics in Maxwell's background, see Wilson, "Educational Matrix," 33, 48n.

20 Wilson, "Educational Matrix," 15–19, on 16; 34, 40–1. Cf. William Whewell's reference to "the two great sciences, Physical Astronomy and Physical Optics" – Larry Laudan, "The Medium and Its Message: A Study of Some Philosophical Controversies about Ether," in Cantor and Hodge, eds., *Conceptions of Ether,* 157–86, on 178.

21 David B. Wilson, "Experimentalists among the Mathematicians: Physics in the Cambridge Natural Sciences Tripos, 1851–1900," *Historical Studies in the Physical and Biological Sciences, 12* (1982), 325–71, quotation from "Report of the Board of Mathematical Studies," 3 June 1850, on 337. Idem, "Educational Matrix," 16. Becher, "Cambridge Mathematics," 23–4, 39, 44.

22 Cf. Becher, "Cambridge Mathematics," 48, and G. N. Cantor, "The Reception of the Wave Theory of Light in Britain: A Case Study Illustrating the Role of Methodology in Scientific Debate," *Historical Studies in the Physical and Biological Sciences, 6* (1975), 109–32, esp. 113.

23 James D. Forbes, "Dissertation Sixth: Exhibiting a General View of the Progress of Mathematical and Physical Science, Principally from 1775 to 1850," in *Encyclopaedia Britannica,* 8th ed. (Edinburgh: Adam & Charles Black, 1855–60), *1,* 795–996, on 805; on 804, also in Balfour Stewart, transcr., "[In Stewart's hand:] Abridgement of Prof[essor] Forbes['s] lectures on Natural Philosophy [in Edinburgh University, 1845–6]," by James David Forbes, Edinburgh University Library, DC.7.101-7, Lecture 3.

24 Stewart, transcr., "Forbes's Lectures," Lecture 3 (perhaps verbatim).

25 Forbes, "Physical Science," 804, where Forbes also referred to the area covered as "Physico-Mathematical Sciences" and cited Whewell for the taxonomy of the sciences.

26 Forbes, "Physical Science," 803–6, on 804; 958–92, on 977. Wilson, "Educational Matrix," 21–4.

27 See Thomas S. Kuhn, *The Structure of Scientific Revolutions*, 2nd. enl. ed. (University of Chicago Press, 1970), 182, for the "disciplinary matrix." Forbes himself had, in a letter to Whewell, 8 August 1833, envisioned the eventual extension of the Cambridge mixed mathematics approach to electricity and magnetism: "Any doubt as to the propriety of viewing mixed mathematics as belonging to a natural philosophy class is at this moment peculiarly untenable; for the whole progress of general physics is happily so fast tending to a subjection to mathematical laws of that department of science, that in no very long time, magnetism, electricity, and light may be expected to be as fully the objects of dynamical reasoning as gravitation is at this present time." [Crosbie Smith, "'Mechanical Philosophy' and the Emergence of Physics in Britain: 1800–1850," *Annals of Science*, *33* (1976), 3–29, on 27]. Forbes's early enthusiasm for the incorporation of Cambridge mixed mathematics into the natural philosophy curriculum had cooled by the time Maxwell attended Forbes's class in 1847–8 and 1848–9 (Wilson, "Educational Matrix," 23), and Forbes's antipathy toward deep theory (see Section 3 in this Chapter) was a countervailing force at that later time; the opinions expressed in this letter of 1833, therefore, may not have had any direct influence on Maxwell.

28 The melding of Scottish and Cambridge traditions in the work of Maxwell, William Thomson, Peter Guthrie Tait, and others has received much attention. See, e.g., Richard Olson, *Scottish Philosophy and British Physics, 1750–1880: A Study in the Foundations of the Victorian Scientific Style* (Princeton University Press, 1975), esp. 287–335; Becher, "Cambridge Mathematics," esp. 46–7; Wilson, "Educational Matrix," esp. 33–45; Harman, "Edinburgh Philosophy and Cambridge Physics"; Smith, "'Mechanical Philosophy.'" Wilson, of these, most clearly shares an emphasis on the importance for Maxwell of the broad Scottish subject matter, as contrasted with the narrower "Cambridge subjects" (p. 41).

29 Wilson, "Educational Matrix," 26–33, 43–5. Jed Z. Buchwald, "Sir William Thomson (Baron Kelvin of Largs)," in *Dictionary of Scientific Biography, 13* (1976), 374–88, esp. 374–7.

30 William Thomson, "On the Uniform Motion of Heat in Homogeneous Solid Bodies and Its Connection with the Mathematical Theory of Electricity" (1842), and "On the Elementary Laws of Statical Electricity" (1845), in *Reprint of Papers on Electrostatics and Magnetism*, 2nd ed. (London: Macmillan, 1884), 1–14 and 15–37. Siegel, "Universal Ether," esp. 240–2.

31 Thomson, "Elementary Laws," 37: "The commonly received ideas of attraction and repulsion exercised at a distance, independently of any intervening medium, are quite consistent with all the phenomena of electrical action which have been here adduced. . . . It is, no doubt, quite possible that such forces at a distance may be discovered to be produced entirely by the action of contiguous particles of some intervening medium, and we have an analogy for this in the case of heat." Cf. Ole Knudsen, "Mathematics and Physical Reality in William Thomson's Electromagnetic Theory," in Harman, ed., *Wranglers and Physicists*, 149–79, esp.

149–57; and M. Norton Wise and Crosbie Smith, "Measurement, Work and In-
dustry in Lord Kelvin's Britain," *Historical Studies in the Physical and Biological
Sciences, 17* (1986), 147–74, esp. 148–51. [Crosbie Smith and M. Norton Wise,
Energy and Empire: A Biographical Study of Lord Kelvin (Cambridge University
Press, 1989), would have enriched the present work, had it come to hand sooner;
as it was, I relied instead upon the earlier works of these authors.]

32 Knudsen, "Thomson's Electromagnetic Theory," 157–64, 167. Sylvanus P. Thompson, *The Life of William Thomson, Baron Kelvin of Largs,* 2 vols. (London: Macmillan, 1910), *1,* 325–96 (Ch. 8, "The Atlantic Telegraph: Failure"), and 397–446
(Ch. 9, "Strenuous Years"), dealing with the years 1854–66.

33 Becher, "Cambridge Mathematics," passim, esp. 6–7, 44. Wilson, "Educational
Matrix," 15. Idem, "Physics in the Natural Sciences Tripos," 334, 367. On the
"mechanical philosophy," in the Scottish context, see Smith, "'Mechanical Philosophy,'" esp. 3–13.

34 Wilson, "Educational Matrix," on 16 (emphasis mine); idem, "Physics in the Natural Sciences Tripos," 334, 367–8.

35 William Whewell, *The Philosophy of the Inductive Sciences, Founded upon Their
History,* 2nd ed., 2 vols. (London: John W. Parker, 1847), inductive tables following *2,* 118. In the case of optics, the reduction was not as fully accomplished as
in celestial mechanics, and the analogy with sound was used "for supporting weak
portions of the mechanical part of the theory," thereby allowing for mechanical
treatment and supporting the program of complete reduction – Wilson, "Educational Matrix," 16–18, on 16; also idem, "The Reception of the Wave Theory of
Light by Cambridge Physicists (1820–1850): A Case Study in the Nineteenth-
Century Mechanical Philosophy," dissertation, Johns Hopkins University, 1968
(University Microfilms No. 68-16,501); and idem, "Concepts of Physical Nature:
John Herschel to Karl Pearson," in U. C. Knoepflmacher and G. B. Tennyson,
eds., *Nature and the Victorian Imagination* (Berkeley: University of California
Press, 1977), 102–215, esp. 202.

36 R. V. Jones, "James Clerk Maxwell at Aberdeen, 1856–1860," *Notes and Records of the Royal Society of London, 28* (1973), 57–81, on 73, from Maxwell's
inaugural lecture at Aberdeen, 3 November 1856. Maxwell reused much of the
material from his Aberdeen inaugural – including the quoted material in unaltered
form – four years later in his inaugural lecture at King's College, London (University Library, Cambridge, Add. MSS 7655). See also Chapter 2.

37 Wilson, "Reception of the Wave Theory," esp. 11–25. Cantor, "Reception of the
Wave Theory."

38 The character of deep theory – the reflective distance between its foundational assumptions and the observational data to which it applies – is evoked in Albert
Einstein, *Ideas and Opinions* (New York: Crown, 1954), based on *Mein Weltbild,*
ed. Carl Seelig, and other sources, "The Fundaments of Theoretical Physics"
(1940), 323–35, esp. 323–5, and "Physics and Reality" (1936), 290–323, esp.
293–5 and 322: ". . . the logical basis departs more and more from the facts of
experience, and . . . the path of our thought from the fundamental basis to those
derived propositions, which correlate with sense experiences, becomes continually
harder and longer."

39 Isaac Newton, *Sir Isaac Newton's Mathematical Principles of Natural Philosophy*

and His System of the World, tr. Andrew Motte/rev. Florian Cajori (Berkeley: University of California Press, 1946), xviii; cf., e.g., Whewell's echo of Newton's statement, "It is probable that almost all the phenomena in the different departments of Natural Philosophy, consist in the insensibly small motions of particles" (Smith, "'Mechanical Philosophy,'" 23), and Maxwell's statement on mechanical reduction in Section 2 of this chapter. On Newton's transdictive leap, see Williams, *Origins of Field Theory*, 32–9. There was, at Cambridge, some disagreement on the general issue of Newton's "analogy of nature" and the properties – such as hardness, extension, impenetrability – of the ultimate constituents of matter. Newton, *Mathematical Principles of Natural Philosophy*, 398; P. M. Harman, *Metaphysics and Natural Philosophy: The Problem of Substance in Classical Physics* (Brighton: Harvester Press; Totowa, N.J.: Barnes & Noble, 1982), 140–5, esp. discussion of Whewell's reservations. Concerning, however, the universal applicability (including to the microcosm) of the Newtonian laws of motion, there was no such disagreement; see the discussion of Herschel and Whewell that follows in the text, as well as David B. Wilson, "Herschel and Whewell's Version of Newtonianism," *Journal of the History of Ideas*, 35 (1974), 79–97, esp. 90–1, and idem, *Kelvin and Stokes: A Comparative Study in Victorian Physics* (Bristol: Adam Hilger, 1987), 101.

40 Cf. Cantor, "Reception of the Wave Theory," 112–13; Jack Morrell and Arnold Thackray, *Gentlemen of Science: Early Years of the British Association for the Advancement of Science* (Oxford: Clarendon Press, 1981), 479–84, esp. 483; Smith, "'Mechanical Philosophy,'" esp. 22–4.

41 William Hopkins, as quoted in Wilson, "Educational Matrix," 16.

42 A pioneering study of the approaches of Herschel and Whewell, especially as applied to the undulatory theory of light, is by Wilson, "Reception of the Wave Theory," esp. 87–131; in Wilson's more recent "Herschel and Whewell's Version of Newtonianism," esp. 94–97, Herschel and Whewell's confidence in this kind of deep theory is related to theological considerations.

43 John F. W. Herschel, *A Preliminary Discourse on the Study of Natural Philosophy*, with a new foreword by Arthur Fine (1830; University of Chicago Press, 1987) [facsimile of the 1830 edition published as vol. I of Dionysius Lardner's *Cabinet Cyclopaedia* (1832)], 191.

44 Ibid., 191, 196–9, 203, 208–9 (emphasis Herschel's). The history of the hypothetico-deductive approach through Herschel and Whewell, especially as applied to the ether, is sketched in Laudan, "The Medium and Its Message."

45 Herschel, *Preliminary Discourse*, 196, 204 (emphasis Herschel's).

46 Wilson, "Reception of the Wave Theory."

47 Laudan, "The Medium and Its Message," 174–8; Whewell, *Philosophy of the Inductive Sciences*, 2, 41–95, where Whewell refers to the process as inductive, but includes in it the use of hypotheses and deductions from them (see esp. 90–3), thus describing an essentially hypothetico-deductive process. Cf. also Robert E. Butts, ed., *William Whewell's Theory of Scientific Method* (University of Pittsburgh Press, 1968), 3–29 (editor's introduction), esp. 17: "His [Whewell's] theory of induction, in its full form, expresses what is now called the hypothetico-deductive character of well-developed sciences."

48 Wilson, "Reception of the Wave Theory," 92–4, 101–6; Larry Laudan, *Science*

and Hypothesis (Dordrecht: Reidel, 1981), 172. Whewell summed up his theory of gradually dawning intuitive certainty epigrammatically in his *On the Philosophy of Discovery* (London: John W. Parker, 1860), 344: *"There are scientific truths which are seen by intuition, but this intuition is progressive"* (emphasis Whewell's).

49 Laudan, *Science and Hypothesis*, 163–80; Whewell, *Philosophy of the Inductive Sciences*, 2, 74–95. Wilson, "Reception of the Wave Theory," 159. On Whewell's later reservations, however, see David B. Wilson, "Convergence: Metaphysical Pleasure Versus Physical Constraint," 44pp. typescript, esp. 28–33, to appear in Menachem Fisch and Simon Schaffer, eds., *William Whewell: A Composite Portrait* (Oxford University Press, in press), 233–54.

50 Whewell, *Philosophy of the Inductive Sciences*, 2, 90, and *Philosophy of Discovery*, 344. Cf. Butts, *Whewell's Theory*, editor's introduction, esp. 21–4. A fuller consideration of Whewell's views would give attention to some crosscurrents and tensions, as in Wilson, "Convergence": On the one hand, Whewell wrote concerning consilient theories that "we feel this irresistible conviction that they are truly & rightly deduced from nature," "this conviction" in the best cases being "accompanied with a glow and exultation" (pp. 7–8); on the other hand, Whewell later had some doubts concerning specific unifying theories that were being proposed – "where others supported convergence, he hesitated" (pp. 28–35, on 32). Cf. discussion of the narrowness of the Cambridge disciplinary matrix in Section 2 of this chapter.

51 In general, on Maxwell's relationship with Whewell and the associated methodological influence, see Harman, "Edinburgh Philosophy and Cambridge Physics," 203, 213, 221–3; idem, *Letters and Papers of Maxwell, 1*, 7–12; Everitt, *Maxwell*, 51–2, 116; Paul Theerman, "Maxwell and Method: Cultural Resonances in the Philosophy of Physics" (paper delivered at the annual meeting of the History of Science Society, Norwalk, Connecticut, 27–30 October 1983). Balfour Stewart's notes on Forbes's natural philosophy lectures of 1845–6 – two years before Maxwell entered the class – indicate that "Dr. Hewell's [*sic*] history & philosophy of the inductive sciences" was recommended and that there was an essay competition bearing on the Mill–Whewell debate – Stewart, transcr., "Forbes's Lectures," Lecture 21. Maxwell referred explicitly to Whewell's methodological ideas in letters to R. B. Litchfield of June 1855, and especially July 1856 (Campbell and Garnett, *Life*, 215, 261), where Maxwell identified with the Whewellian (indeed ultra-Whewellian) camp: "I find I get fonder of metaphysics and less of calculation continually, and my metaphysics are fast settling into the rigid high style, that is about ten times as far *above* Whewell as Mill is *below* him . . . using above and below conventionally." Maxwell went on in the letter to develop a line of thought reflecting Whewell's notion of progressive intuition leading eventually to certainty. For a balanced statement of Maxwell's appreciation of Whewell, dating from a later period, see James Clerk Maxwell, "Whewell's Writings and Correspondence" (1876), in Niven, ed., *Scientific Papers of Maxwell, 2*, 528–32.

52 Wilson, "Educational Matrix," 23–4; idem, "Concepts of Physical Nature," 207–8; Stewart, transcr., "Forbes's Lectures," Lecture 94, see also Lecture 21; Campbell and Garnett, *Life*, on 302, 305, 359; Everitt, *Maxwell*, 64, 136–7; Maxwell,

"On the Theory of Colours in Relation to Colour-Blindness" (1855) and "Experiments on Colour, as Perceived by the Eye, with Remarks on Colour-Blindness," in Niven, ed., *Scientific Papers of Maxwell, 1,* 119–25, esp. 119, and 126–54, esp. 142.

53 Herschel, *Preliminary Discourse,* 204, 191. *The Works of Francis Bacon,* eds. James Spedding, Robert Leslie Ellis, and Douglas Denon Heath, new ed. (London: Longmans, 1857–74; New York: Garrett Press, 1968), *1,* 204–5; *4,* 97 (English translation, *The New Organon,* "Aphorisms Concerning the Interpretation of Nature and the Kingdom of Man," Book I, CIV).

54 Forbes, "Physical Science"; Stewart, transcr., "Forbes's Lectures"; *Works of Francis Bacon, 4,* 127 (*New Organon,* "Aphorisms Concerning the Interpretation of Nature and the Kingdom of Man," Book II, X); Newton, *Mathematical Principle of Natural Philosophy;* René Descartes, "The Principles of Philosophy [Selections]," in *The Philosophical Works of Descartes,* tr. Elizabeth S. Haldane and G. R. T. Ross, 2 vols. (1911; Cambridge University Press, 1970), *1,* 201–302.

55 Forbes, "Physical Science," 974–5; cf. Stewart, transcr., "Forbes's Lectures," Lecture 49, where the Ampèrian hypothesis was characterized as "doubtful." Archibald, "Action-at-a-Distance Electromagnetic Theory."

56 Forbes, "Physical Science," 981, 979; Forbes to Maxwell, 31 March 1857, in Campbell and Garnett, *Life,* 267.

57 Forbes, "Physical Science," 942, 919; see also Smith, "'Mechanical Philosophy,'" 29. On conservation of energy well received as a major organizing principle, see, e.g., John Theodore Merz, *A History of European Thought in the Nineteenth Century,* 4 vols. (1904–12; New York: Dover, 1965), *2,* 133–44.

58 Forbes, "Physical Science," 803. Olson, *Scottish Philosophy and British Physics,* esp. 225–36. William Thomson, "On the Propagation of Laminar Motion through a Turbulently Moving Inviscid Fluid," *Philosophical Magazine, 24* (1887), 342–53, on 352 (emphasis Thomson's).

59 Robert J. McRae, "The Origin of the Conception of the Continuous Spectrum of Heat and Light," dissertation, University of Wisconsin, 1969, 260–368. James D. Forbes, "On the Refraction and Polarization of Heat," *Transactions of the Royal Society of Edinburgh, 13* (1836), 131–68, quoted material on 147n, as cited and discussed in Wilson, "Educational Matrix," 25; cf. Olson, *Scottish Philosophy and British Physics,* 227–9.

60 Olson, *Scottish Philosophy and British Physics,* 229–32; Wilson, "Educational Matrix," on 25; Forbes, "Physical Science," 979.

61 Buchwald, "Sir William Thomson," 377–84.

62 Thomson, "Elementary Laws," 37.

63 M. Norton Wise, "William Thomson's Mathematical Route to Energy Conservation: A Case Study of the Role of Mathematics in Concept Formation," *Historical Studies in the Physical and Biological Sciences, 10* (1979), 49–84, esp. 62.

64 Donald Franklin Moyer, "Energy, Dynamics, Hidden Machinery: Rankine, Thomson and Tait, Maxwell," *Studies in History and Philosophy of Science, 8* (1977), 251–68.

2. Mechanical image and reality in Maxwell's electromagnetic theory

1 Maxwell, "Faraday's Lines," 156. See further Maxwell's "Are there Real Analogies in Nature?" an essay presented to the Cambridge Apostles' Club in February

1856, in Campbell and Garnett, *Life*, 235–44; see also the discussion in Harman, "Edinburgh Philosophy and Cambridge Physics," esp. 213. Further, on the background to "Faraday's Lines," see Harman, ed., *Letters and Papers of Maxwell*, *1*, 12–15.

2   Maxwell to Thomson, 15 May 1855, in Larmor, ed., "Familiar Letters," 705; also in Harman, ed., *Letters and Papers of Maxwell*, *1*, 307.

3   Maxwell, "Faraday's Lines," 155–9.

4   Ibid., 155–229, esp. 155–60. On the use of the word "theory" to denote a comprehensive scheme, associated with "the higher degrees of inductive generalization," see Herschel, *Preliminary Discourse*, 190. The interactions among electric fields, magnetic fields, and electric currents were characterized mathematically – without the aid of the flow analogy, which could render no service in this connection – in Part II of "Faraday's Lines," 188–209; see Everitt, "Maxwell's Creativity," 120–6.

5   Maxwell, "Faraday's Lines," 207–8.

6   Ibid., 157, 159.

7   On the term "physical" as "commonly employed" to denote "the . . . portion . . . of [a] science" that goes beyond description to deal with "causes" – such as the "forces by which the heavenly bodies are guided" in "Physical Astronomy," or "the machinery by which the effects are produced" in "Physical Optics" – see Whewell, *Philosophy of the Inductive Sciences*, *2*, 96–7. Ludwig Boltzmann, ed. and tr., *Ueber Faraday's Kraftlinien*, Ostwald's Klassiker der exacten Wissenschaften, No. 69 (Leipzig: Wilhelm Engelmann, 1895), 9, translates "a mature theory, in which physical facts will be physically explained," as "einer definitiven Theorie, welche die physikalischen Tatsachen durch bestimmte Annahmen über das Wesen der Dinge erklärt," or "a definitive theory, in which physical facts will be explained on the basis of definite suppositions concerning the nature of things." On the connection of physical explanation with mechanical reduction for Maxwell, see Chapter 1, Section 2.

8   On Maxwell's accomplishments in electromagnetic theory in the context of the analogical approach in "Faraday's Lines" (which included the hoped-for "simplification and reduction of the results of previous investigation to a form in which the mind can grasp them," eventuating in field quantities and equations that we shall take as our starting point in the following chapters), see M. Norton Wise, "The Flow Analogy to Electricity and Magnetism," dissertation, Princeton University, 1977 (University Microfilms No. 77-14,252), and the sequelae, "The Flow Analogy to Electricity and Magnetism. Part I: William Thomson's Reformulation of Action at a Distance," *Archive for History of Exact Sciences*, *25* (1981), 19–70, esp. 67–70, and "The Flow Analogy to Electricity and Magnetism. Part II: Maxwell's First Formulation of Field Theory," (in press); also Everitt, *Maxwell*, 87–93; Rosenfeld, "Evolution of Electrodynamics," esp. 1652–5.

9   Thompson, *Life of Kelvin*, *1*, 203.

10 William Thomson, "Dynamical Illustrations of the Magnetic and Helicoidal Rotatory Effects of Transparent Bodies on Polarized Light," *Proceedings of the Royal Society*, *8* (1856), 150–8, on 152; reprinted in *Philosophical Magazine*, *13* (1857), 198–204.

11 Wise, "Thomson's Mathematical Route," 62. Thomson, *Electrostatics and Magnetism*, 423 ("Note added Jan. 1872").

12   William Thomson, "On an Absolute Thermometric Scale Founded on Carnot's
     Theory of the Motive Power of Heat, and Calculated from Regnault's Observa-
     tions" (1848); "An Account of Carnot's Theory of the Motive Power of Heat;
     with Numerical Results Deduced from Regnault's Experiments on Steam" (1849);
     "On the Dynamical Theory of Heat, with Numerical Results Deduced from Mr.
     Joule's Equivalent of a Thermal Unit, and M. Regnault's Observations on Steam"
     (1851); all in idem, *Mathematical and Physical Papers*, 6 vols. (Cambridge Uni-
     versity Press, 1882–1911), *1*, respectively 100–6, esp. 102–3; 113–55, esp. 116–
     17; and 174–232. Martin J. Klein, "Gibbs on Clausius," *Historical Studies in the
     Physical and Biological Sciences*, *1* (1969), 127–49. Thompson, *Life of Kelvin*,
     1, 263–83.
13   William Thomson, "On a Universal Tendency in Nature to the Dissipation of Me-
     chanical Energy" (1852), in *Mathematical and Physical Papers*, *1*, 511–14, on
     511; idem, "Dynamical Theory of Heat," 174–5; idem, *Electrostatics and Magne-
     tism*, 423 ("Note added Jan. 1872"); idem, "Dynamics," in [*Nicol's*] *Cyclopaedia
     of the Physical Sciences* (London: Richard Griffin, 1860), 212–15.
14   Thomson, "Dynamical Illustrations," on 199–200. Idem, "Dynamical Theory of
     Heat," 174–5; William John Macquorn Rankine, "On the Centrifugal Theory of
     Elasticity, as Applied to Gases and Vapours," in *Miscellaneous Scientific Papers*,
     ed. W. J. Millar (London: Charles Griffin & Company, 1881), 16–48, esp. 16–
     18. Joule at first espoused the hypothesis that heat was rotatory motion, but then
     gave it up. See Stephen G. Brush, *The Kind of Motion We Call Heat*, 2 vols.
     (New York: North-Holland, 1976), *1*, 161.
15   Michael Faraday, "On the Magnetization of Light and the Illumination of Magnet-
     ic Lines of Force" (1845), in *Experimental Researches*, *3*, 1–26.
16   Thomson, "Dynamical Illustrations," 151–2.
17   William Thomson, "Atmospheric Electricity (Royal Institution Friday evening lec-
     ture, May 18, 1860)," in *Electrostatics and Magnetism*, 208–26, on 224 (Thom-
     son's emphasis).
18   Thomson, "Dynamical Illustrations," 152.
19   Maxwell, "Physical Lines," 505; idem, "Faraday's lines," 159. Wilson, "Concepts
     of Physical Nature," 207–8 (see also the works cited there), sees "Victorian real-
     ism" as theologically based and argues this for the cases of Herschel, Whewell,
     and Thomson; Wilson further suggests that Maxwell would have been a realist for
     the same reasons. See further Wilson, *Kelvin and Stokes*, 100–28 (chapter entitled
     "Cautious Realism").
20   As recalled by Lewis Campbell, Maxwell had, probably in 1857, "described to
     [him], with extraordinary volubility, the swift, invisible motions by which mag-
     netic and galvanic phenomena were to be accounted for" in the theory of molecu-
     lar vortices. Campbell and Garnett, *Life*, 199n.
21   Maxwell to Cecil James Monro, 20 May 1857, in Harman, ed., *Letters and Pa-
     pers of Maxwell*, *1*, 507; Maxwell to Faraday, 9 November 1857, in Campbell
     and Garnett, *Life*, xvii, also in Harman, ed., *Letters and Papers of Maxwell*, *1*,
     552; Maxwell to Thomson, 30 January 1858, in Stephen G. Brush, C. W. F.
     Everitt, and Elizabeth Garber, eds., *Maxwell on Saturn's Rings* (Cambridge,
     Mass.: MIT Press, 1983), 61, also in Harman, ed., *Letters and Papers of Max-
     well*, *1*, 579–80.

22 Maxwell to Thomson, 24 November 1857, in Harman, ed., *Letters and Papers of Maxwell, 1*, 560–3. On Herschel and Whewell, see Chapter 1, Section 3; further, on Thomson's thinking at that time concerning the possible explanation of "all the phenomena of matter" on the basis of vortices in "the Universal Plenum," see Ole Knudsen, "From Lord Kelvin's Notebook: Ether Speculations," *Centaurus, 16* (1972), 41–53, on 47; see also Siegel, "Universal Ether," esp. 245–6, 254–7. The long gestation period for "Physical Lines" is emphasized in Harman, ed., *Letters and Papers of Maxwell, 1*, 30–1; on Maxwell's thought processes during such gestation periods, see Everitt, "Maxwell's Creativity," 132–4.

23 Maxwell, "Physical Lines," 451–513, on 452. Cf. Herschel, *Preliminary Discourse, 204*.

24 Faraday, "On the Physical Character of the Lines of Magnetic Force" (1852), in *Experimental Researches, 3*, 407–37, esp. 419, 435–7; on 437 (emphasis Faraday's). Faraday's usage of the term "physical" accords well with Whewell's usage, and Faraday (and others) often looked to Whewell as an expert on appropriate terminology; neither Faraday nor Maxwell, however, picked up on Whewell's use of the term "formal" as opposed to "physical." Whewell, *Philosophy of the Inductive Sciences, 2*, 96–7; Williams, *Faraday, 259–66, 382.*

25 Maxwell, "Faraday's Lines," 159, and "Physical Lines," 452–5, 467 (emphasis mine). On the goal of explanation, in terms of real causes, of the Cambridge hypothetico-deductive enterprise, see Herschel, *Preliminary Discourse*, 144ff., 197, and Whewell, *Philosophy of the Inductive Sciences, 95–106.*

26 For a broader discussion of theories and analogies, see, e.g., Achinstein, *Concepts of Science*, esp. 203–25; my emphasis on forces as the *explicanda* is directed to the specific issues at hand, making my discussion less than completely general.

27 Maxwell, "Physical Lines," 451–2. Cf. Knudsen, "Faraday Effect," 248–52, for a judgment similar to that made here, namely, that for Maxwell, "a theory which is both mechanical and true is something infinitely superior to [and very different from] a physical analogy," and that by saying, in "Physical Lines," that he was proposing a "theory," Maxwell was indicating that he "had now found the true mechanical theory which he in 1855 could only hope for." (Cf., however, Knudsen, "Thomson's Electromagnetic Theory," 166–7, for a varying nuance.) For parallel conclusions, to the effect that the "task" (Turner) or "programme" (Harman) of "Physical Lines" was indeed "set" or "foreshadowed" in "Faraday's Lines," see Harman, "Consistent Representation," 188, and Joseph Turner, "A Note on Maxwell's Interpretation of Some Attempts at Dynamical Explanation," *Annals of Science, 11* (1955), 238–45, on 241.

28 Maxwell, "Physical Lines," 468, 485, 485n; see also Appendix 1.

29 Maxwell to Faraday, 19 October 1861, in Campbell and Garnett, *Life*, xx–xxii, on xxii; also in Harman, ed., *Letters and Papers of Maxwell, 1*, 683–9, on 688, photograph of the apparatus opposite. Maxwell wrote in similar terms to his other mentor in electricity and magnetism – Maxwell to Thomson, 10 December 1861, in Larmor, ed., "Familiar Letters," 728–30, on 730; also in Harman, ed., *Letters and Papers of Maxwell, 1*, 692–8, on 698 – where Maxwell briefly mentioned the experiment and complained that "I . . . have not yet got rid of the effects of terrestrial magnetism which are very strong on a powerful electromagnet which I

use." A week later – Maxwell to Thomson, 17 December 1861, in Larmor, ed., "Familiar Letters," 731–3, on 733; also in Harman, ed., *Letters and Papers of Maxwell, 1,* 699–702, on 702 – Maxwell reiterated his determination to find "the actual breadth" of "a vortex of magnetism" by experiment.

30 Maxwell, *Treatise, 2,* 202–4 (1st ed.), 218–20 (3rd ed.); Everitt, *Maxwell,* 106–8.

31 Maxwell, "Physical Lines," 468–71, 486, and Plate VIII facing 488. Finding the phrase "molecule of electricity" useful in a discussion of electrolysis – idem, *Treatise, 1,* 312 (1st ed.), 380 (3rd ed.) – Maxwell noted that "this phrase, gross as it is, and out of harmony with the rest of this treatise," nevertheless had heuristic value. Herschel, *Preliminary Discourse,* compare 196 and 203–4, 208.

32 Maxwell, "Physical Lines," 488; Maxwell to Thomson, 10 December 1861, in Larmor, ed., "Familiar Letters," 728–9; Everitt, *Maxwell,* 98–9. On the psychological aspects of disappointment and "intellectual pain" stimulating rather than weakening Maxwell's "tenacity" in intellectual tasks, see Everitt, "Maxwell's Creativity," 110–18, on 118.

33 Maxwell, "Physical Lines," 489–90, 497–500.

34 Ibid., 489–90; Maxwell to Faraday, 19 October 1861, in Campbell and Garnett, *Life,* xx–xxii, on xx; Maxwell to Thomson, 10 December 1861, in Larmor, ed., "Familiar Letters," 728–30. Cf. Maxwell, *Treatise, 2,* 383 (1st ed.), 431 (3rd ed.); Bromberg, "Displacement Current," 227–9. Laudan, "The Medium and Its Message," 173–81.

35 Olson, *Scottish Philosophy and British Physics,* 302–3.

36 Maxwell, "Physical Lines," 502–3 (emphasis Maxwell's).

37 James Clerk Maxwell, "On the Mathematical Classification of Physical Quantities," (1870), in Niven, ed., *Scientific Papers of Maxwell, 2,* 257–66, esp. 263; cf., e.g., Philip M. Morse and Herman Feshbach, *Methods of Theoretical Physics* (New York: McGraw-Hill, 1953), 1–44. Everitt, *Maxwell,* 123–4.

38 Jones, "Maxwell at Aberdeen," 73, from Maxwell's inaugural lecture at Aberdeen; see also Chapter 1, Section 2.

39 Maxwell, "Physical Lines," 503–4.

40 Ibid., 504–6.

41 Ibid., 505–6.

42 Ibid., 505. This is not to say that Maxwell was necessarily a naive realist in a philosophical sense; his "Are there Real Analogies in Nature" in fact suggests leanings away from naive realism and toward the view that all knowledge is analogical and relational. The context there was the question of the relationship between "the constitution of the intellect and that of the external world," as regards notions of space, time, number, and cause, and as generalizable to moral propositions. Campbell and Garnett, *Life,* 235–44, on 238. The issue, then, was of the status of all thought and knowledge vis-à-vis the external world, and Maxwell's opinions on this subject, though not particularly transparent, were certainly not naive or uninformed with respect to philosophical alternatives. Thus, when Maxwell referred to "a real rotation going on in the magnetic field," one presumes he meant that this rotation was as real as the rotation of a directly observed macroscopic spinning top; he did not necessarily mean to say that our mental image of either spinning object was anything more than an analogy to the referent in the "external world."

43 James Clerk Maxwell, "A Dynamical Theory of the Electromagnetic Field" (read
    8 December 1864, pub. 1865), in Niven, ed., *Scientific Papers of Maxwell, 1*,
    526–97.
44 Maxwell to Thomson, 24 November 1857, in Harman, ed., *Letters and Papers of
    Maxwell, 1*, 560–3; see also the discussion in Section 2 of this chapter. James
    Clerk Maxwell, "Illustrations of the Dynamical Theory of Gases" (1860), in
    Niven, ed., *Scientific Papers of Maxwell, 1*, 377–409; also Maxwell to Stokes,
    30 May 1859, in Elizabeth Garber, Stephen G. Brush, and C. W. F. Everitt, eds.,
    Maxwell on Molecules and Gases (Cambridge, Mass.: MIT Press, 1986), 277–81;
    Stephen G. Brush, ed., *Kinetic Theory. Vol. 1: The Nature of Gases and of Heat*
    (Oxford: Pergamon Press, 1965), 22–30; Daniel M. Siegel, "The Energy Con-
    cept: A Historical Overview," *Materials and Society, 7* (1983), 411–24, esp. 416.
    Cf. a similar instance of preferring the rotatory theory – "the beautiful hypothesis
    of a rotary motion" – while employing the linear picture for purposes of calcula-
    tion because it "is somewhat simpler," in James Prescott Joule, "Some Remarks
    on Heat and the Constitution of Elastic Fluids" (1848), in *The Scientific Papers of
    James Prescott Joule* (London: Taylor & Francis, 1884), 290–7, on 293, 294.
45 James Clerk Maxwell, "On the Dynamical Theory of Gases" (1866), in Niven,
    ed., *Scientific Papers of Maxwell, 2*, 26–78, esp. 27; Everitt, *Maxwell*, 95. Con-
    cerning methodological parallels between Maxwell's work on gas theory and his
    work in electromagnetic theory, see also Garber, Brush, and Everitt, eds., *Max-
    well on Molecules*, xxii–xxv.
46 Bromberg, "Displacement Current," 219; cf. C. W. F. Everitt, "Maxwell's Scien-
    tific Papers," *Applied Optics, 6* (1967), 639–46, esp. 643.
47 Maxwell to C. Hockin, 7 September 1864, in Campbell and Garnett, *Life*, 340;
    cf. C. J. Monro to Maxwell, 23 October 1861, idem, p. 329; Maxwell, "Dy-
    namical Theory," section entitled "Electromagnetic Theory of Light," 577–88, see
    also James Clerk Maxwell, "On a Method of Making a Direct Comparison of
    Electrostatic With Electromagnetic Force; With a Note on the Electromagnetic
    Theory of Light" (1868), in Niven, ed., *Scientific Papers of Maxwell, 2*, 125–43.
    See, further, Everitt, *Maxwell*, 99–101; Bromberg, "Displacement Current," 227–
    30. Also see Chapters 5–6 herein.
48 Maxwell, "Dynamical Theory," 529–30, 564. Maxwell did make use of a mac-
    roscopic mechanical analogy for inductive circuits that was in some sense – al-
    though not the usual sense – a mechanical representation, beginning in
    "Dynamical Theory" and fully developed later (Everitt, *Maxwell*, 102–5).
      Maxwell's "dynamical theory" cannot be reduced to some combination of the
    Scottish, Cambridge, and Thomsonian approaches with which he had started out.
    Instead, while having some precedent in certain kinds of optical ether theory, "dy-
    namical theory," in its mature development, was an emergent feature of Max-
    well's research program and his interaction with William Thomson and Peter
    Guthrie Tait in the 1860s and early 1870s. A full discussion of that fruitful inter-
    action and the attendant development of the notion of dynamical theory would
    take us far from our focus on the molecular-vortex model; see further, on this top-
    ic, Siegel, "Universal Ether," 259–63, and the works cited there; cf. Wise, "Brit-
    ish Dynamical Theory," and the works cited there. More recent on the topic are
    Everitt, "Maxwell's Creativity," 128–9; Harman, "Newton to Maxwell"; and

idem, "Mathematics and Reality in Maxwell's Dynamical Physics," in Robert Kargon and Peter Achinstein, eds., *Kelvin's Baltimore Lectures and Modern Theoretical Physics: Historical and Philosophical Perspectives* (Cambridge, Mass.: MIT Press, 1987), 267–97.

49   Maxwell, *Treatise, 2,* 399–408, on 406, 408 (1st ed.); 451–61, on 459, 461 (3rd ed.). Cf. a postcard from Maxwell to Thomson, undated but probably from that period, in Larmor, ed., "Familiar Letters," 748; see, in addition, a much condensed version of the redundancy argument of Part IV of "Physical Lines," in "Mathematical Classification of Physical Quantities," 263. Cf. Knudsen "Faraday Effect," esp. 273–81, and Bromberg, "Maxwell's Electrostatics," 145.

50   Maxwell, "Physical Lines," 468, 486; idem, *Treatise, 2,* 415–17 (1st ed.), 468–70 (3rd ed.).

51   Maxwell, *Treatise, 2,* 416–17 (1st ed.), 470 (3rd ed.).

52   Ibid. Cf. the discussion of "imaginary models" and the contrast with "analogue models" in Achinstein, *Concepts of Science,* esp. 218–25; cf. also Knudsen, "Thomson's Electromagnetic Theory," 166–7, contrast of "dynamical illustration" with "mathematical analogy." Cf. also Herschel, *Preliminary Discourse,* 204, on the possibility of many models – many "mechanism[s] . . . leading to the same results."

53   On one occasion Maxwell did in fact evaluate the whole theory by its weakest link, the connecting mechanism. In a letter to P. G. Tait of 23 December 1867, Maxwell wrote that the "vortex theory . . . is built up to show that the phenomena are such as can be explained by mechanism. The nature of the mechanism is to the true mechanism what an orrery is to the Solar System." (Cargill Gilston Knott, *Life and Scientific Work of Peter Guthrie Tait, Supplementing the Two Volumes of Scientific Papers Published in 1898 and 1900* [Cambridge University Press, 1911], 215.) Thus, Maxwell was here characterizing the whole molecular-vortex apparatus as a demonstration model – a "working model," in the language of the *Treatise, 2,* 416–17 (1st ed.), 470 (3rd ed.). This was in a context where Maxwell's rhetorical purpose was not to explore nuances in the status of the molecular-vortex model but rather to emphasize to Tait the difference between the abstract dynamical approach of "Dynamical Theory" and the concrete modeling of "Physical Lines": "The latter [paper, 'Dynamical Theory,'] is built on Lagrange's Dynamical Equations and is not wise about vortices." (Knott, *Life of Tait,* 215.) In this rhetorical context, Maxwell did not distinguish the different parts of the model of "Physical Lines" and their different ontological status.

54   Maxwell, *Treatise, 2,* 416–17 (1st ed.), 470 (3rd ed.).

55   James Clerk Maxwell, "On Action at a Distance" (1873), 311–23, on 321; "Ether" (1879), 763–75, on 774; "Faraday" (1879), 786–93, on 792; all in Niven, ed., *Scientific Papers of Maxwell, 2.* Cf., also therein, idem, "Atom" (1875), 445–84, esp. 466–77, where the "vortex atom" is discussed in most enthusiastic terms; see further Siegel, "Universal Ether," 254–9, and idem, "Mechanical Image and Reality in Maxwell's Electromagnetic Theory," in Harman, ed., *Wranglers and Physicists,* 180–201, esp. 199–200.

56   For a recent and sustained analysis in this vein, see John Hendry, *James Clerk Maxwell and the Theory of the Electromagnetic Field* (Bristol: Adam Hilger, 1986); see also Wise, "British Dynamical Theory," and the works cited there.

57   Ivan Tolstoy, *James Clerk Maxwell: A Biography* (1981; University of Chicago Press, 1982), 78, 124.

3. The elaboration of the molecular-vortex model

1   Everitt, "Maxwell's Creativity," 110–20, on 119, explores psychological reasons for Maxwell's search for "unity" and "coheren[ce]" and characterizes Maxwell as an "architectural thinker"; cf. idem, *Maxwell*, 129–30. On the process of the development and elaboration of theories, cf. Norman Robert Campbell, *Foundations of Science: The Philosophy of Theory and Experiment*, formerly titled *Physics: The Elements* (Cambridge University Press, 1920; New York: Dover, 1957), 132–7 (section entitled "The Development of Theories").
2   Maxwell, "Physical Lines," 451–66; see Chapter 2 herein.
3   Ibid., 453, 457–8, 505–6.
4   Ibid., 505–6; Maxwell to Thomson, 10 December 1861, in Larmor, ed., "Familiar Letters," 728. On Thomson's criticisms of Maxwell's theory, see Knudsen, "Thomson's Electromagnetic Theory," 171–6.
5   Maxwell, "Physical Lines," 452.
6   Knudsen, "Faraday Effect," 261–72.
7   Maxwell, "Physical Lines," 452.
8   Ibid., 454–5.
9   Ibid., 456–8, where Maxwell cited "Rankine's *Applied Mechanics*" for the formalism of stress analysis; Maxwell probably made use of the first, 1858 edition – cf., e.g., William John Macquorn Rankine, *A Manual of Applied Mechanics*, 5th ed. (London: Charles Griffin & Company, 1869), 82–116, to which Maxwell's usage corresponds. For a modern tensor-dyadic representation of stresses in the electromagnetic field, see, e.g., J. D. Jackson, *Classical Electrodynamics* (New York: Wiley, 1962), p. 193. Aspects of the derivation of equation (3.1), including certain questions pertaining to proportionality constants, will be taken up later. Concerning the representation of Maxwell's component equations in vector/tensor form for mnemonic purposes, see the Introduction herein (it is interesting that Rankine does use a matrixlike square array – *Applied Mechanics*, 89).
10   Maxwell, "Physical Lines," 458–9. Maxwell to Thomson, 10 December 1861, in Larmor, ed., "Familiar Letters," 728, where a more analogical tone is employed: "Two similar systems," one electromagnetic and one mechanical, are considered, and correspondences are drawn between the variables characterizing the two systems. Cf. also Campbell, *Foundations of Science*, 119–58, esp. 122–9, where what Campbell calls the "dictionary" sets out something like what I have called *correspondences*.
11   Maxwell, "Physical Lines," 458; cf. Rankine, *Applied Mechanics*, 113–16. It was Maxwell, in 1870 – see his "Mathematical Classification of Physical Quantities," esp. 264–5 – who first suggested the term "*Curl*" for the "vector portion" of the result of the "Hamiltonian operator [del]"; he also suggested "slope" and "*Convergence*" for our **grad** and −div. See further Michael J. Crowe, *A History of Vector Analysis* (University of Notre Dame Press, 1967), esp. 117–39.
12   Cf. the treatment in Maxwell, *Treatise*, 2, 253–8 (1st ed.), 278–83 (3rd ed.), where it is assumed that the properties of magnetic materials can be referred to molecular electric currents: The first term on the right-hand side of equation (3.5)

is then found always to be equal to zero; it is arranged for the equivalents of the second and fourth terms to cancel; and only the term in **curl H** is left.

13 Maxwell, "Physical Lines," 499.

14 Ibid., 456–7, where our correspondence (3.6) is introduced as follows: "In future we shall write $\mu/4\pi$ instead of $C\rho$."

15 Ibid., 492.

16 Though he did not take the matter up explicitly, Maxwell probably realized that there was a free parameter – corresponding to our b – in the theory. His remarks at ibid., 467, where he refers to the relationships expressed in our correspondences (3.2) and (3.3) as proportionalities rather than equalities, is suggestive of this; and at p. 512 he regards the density of the medium as still adjustable, perhaps having something like the parameter b in mind. Maxwell explicitly recognized the existence of another free parameter in the theory, characterizing the sizes of the vortices – see his note, pp. 485–6. See also Maxwell to Faraday, 19 October 1861, in Campbell and Garnett, *Life*, xxii: "The absolute diameter of the magnetic vortices, their velocity and their density . . . as yet . . . are all unknown." Cf. the discussion in Chapter 2, Section 3.

17 Maxwell, "Physical Lines," 467–8.

18 Ibid., 467–8, 486; Maxwell to Thomson, 10 December 1861, in Larmor, ed., "Familiar Letters," on 728. See also Chapter 2, Section 2, esp. notes 20 and 21.

19 Maxwell, "Physical Lines," 467–9, and Figure 2, Plate VIII, facing 488. Maxwell's description of the wheels as "in gear" has prompted at least one commentator – Martin Goldman, *The Demon in the Aether: The Story of James Clerk Maxwell* (Edinburgh: Paul Harris Publishing, 1983), 150 – to conclude that "Maxwell [had] filled the universe with a complex whirring and clanking aether machine of molecular vortex gears and intermeshed electrical idle wheel cogs" (see illustrations of meshing toothed wheels on p. 148). In nineteenth-century usage, however, "*gearing* and *gear* are the words used to indicate the combination of any number of parts in a machine which are employed for a common object," including but not limited to "toothed wheels" – Thomas Minchin Goodeve, *Elements of Mechanism*, new ed. (London: Longmans, Green & Co., 1892), 22; cf. *Oxford English Dictionary*, s.v. "gear"; Maxwell himself made no mention of teeth or cogs, and his own illustration (Figure 2.2, redrawn herein as Figure 3.1) shows no teeth or cogs; thus, it is a bit fanciful to attribute teeth or cogs to his vortices and idle wheels.

20 Maxwell cited "Goodeve's *Elements of Mechanism*." Cf. Goodeve, *Elements of Mechanism*, 316–18, for the Siemens governor, and 222ff. for epicyclic trains, also mentioned by Maxwell. The examples are somewhat oblique: In the epicyclic train, which is a train of wheelwork or gearing in a movable frame, not only the idle wheel moves; in the Siemens governor, beveled gears are utilized, yielding a differential gearing arrangement.

21 Maxwell, "Physical Lines," 468–71, resulting in equations (33) and (34), on 471.

22 Ibid., 471. This differential form of Ampère's law had appeared before in "Faraday's Lines," 194, and Part I of "Physical Lines," 462. Our ι and $\mathbf{J}$ both correspond to Maxwell's p, q, r, reflecting explicitly his shifting interpretation of this vector; see discussion of ω^* and $\mathbf{H}$ in connection with correspondence (3.2).

23 Maxwell, "Physical Lines," 471. The logic by which Maxwell extended the theo-

ry of molecular vortices has its parallels with the process described in Campbell's fable concerning the development and elaboration of the kinetic theory of gases: *Foundations of Science*, 134–7.

24 Maxwell, "Physical Lines," 469–70, where our **v** is Maxwell's u, $[v, w]$, our **n** is Maxwell's l, m, n, and Maxwell calculates components $n\beta - m\gamma$, $l\gamma - n\alpha$, $m\alpha - l\beta$ for **v**.

25 Ibid., 492ff.; see also Appendix 1 herein.

26 Ibid., Figure 2 of Plate VIII, facing p. 488.

27 Cf. ibid., 486: "the angular velocity must be the same throughout each vortex."

28 Ibid., 471, where our σ is Maxwell's ρ; that σ is measured in electromagnetic units of charge per unit area is not made explicit, but is implied by the complete identification of ι with **J**, and is consistent with all that follows.

29 ι, depending linearly on both ω^* and σ, is not at all affected by the value of b – see equations (3.2') and (3.9'). Cf. the discussion in Chapter 5.

30 Equation (3.7a) can in fact be derived as an exact result for the simple case in which the vortex cells are assumed to be cubical, as in Boltzmann, ed., *Ueber physikalische Kraftlinien*, 108–9, and R. A. R. Tricker, *The Contributions of Faraday and Maxwell to Electrical Science* (Oxford: Pergamon Press, 1966), 115–16; in this case, which involves substantial deviation from sphericity, the behavior of the vortex fluid required by equation (3.8) will be a problem, and neither Boltzmann nor Tricker suggests that Maxwell was actually considering cubical vortices! The derivation does, however, serve to demonstrate the basic reasonableness of Equation (3.7a) as a mechanical result.

31 Maxwell, "Physical Lines," 471.

32 Ibid., 474; τ_1 itself is thus the total force exerted, by two adjacent vortex surfaces, per unit quantity (in electromagnetic units) of idle-wheel particles in the interposed layer.

33 Ibid., 472–5, resulting in Maxwell's equation (54) on p. 475; Maxwell actually treated the numerical coefficient multiplying ρ_m in a somewhat more general way here – cf. the discussion in connection with correspondence (3.6).

34 Ibid., 475–6. On the vector potential, see Alfred M. Bork, "Maxwell and the Vector Potential," *Isis*, 58 (1967), 210–22, and Everitt, *Maxwell*, 97–8, 128–9.

35 Maxwell, "Physical Lines," 475–6. τ_1, depending linearly on ρ_m and ω^* and inversely on σ, is not at all affected by the value of b – see correspondences (3.2') and (3.3') and equation (3.9'). Cf. the discussion in Chapter 5.

36 Maxwell, "Physical Lines," 486.

37 See, e.g., Émile Zola, *The Experimental Novel and Other Essays*, tr. Belle M. Sherman (1893; New York: Haskell House, 1964), 7–9, where the novelist, in parallel with the experimental scientist as described by Claude Bernard, is seen as setting up the initial conditions, then retiring to the position of an "'observer . . . without any preconceived idea.'" Thus, "the novelist . . . sets his characters going in a certain story so as to show that the succession of facts will be such as the requirements of the determinism of the phenomena under examination call for. Here it is nearly always an experiment '*pour voir,*' as Claude Bernard calls it. . . . An experimental novel, 'Cousine Bette,' for example, is simply the report of the experiment that the novelist conducts before the eyes of the public." Material for drawing a parallel between "Physical Lines" and the experimental novel is fur-

nished by Everitt, "Maxwell's Creativity," in its references to "the daring imaginativeness of the molecular vortex model" (p. 83; cf. p. 134), to Maxwell's writing "the paper as an experiment to see how far he could go in accounting by an ether model . . . for electromagnetic phenomena" (p. 127), and to the element of "surprise" in the outcome (p. 127; cf. p. 134).

38   Maxwell, "Physical Lines," 486.

39   Ibid.; cf. the MS fragment "Prop XII" herein, Appendix 1, where alternatives are considered and a choice is made for uniform angular velocity.

40   Ibid., 488; cf. Chapter 2, Section 3, herein.

41   Ibid., 477.

42   For equation (3.11b), see Maxwell, "Faraday's Lines," 191–2, where the symbol ρ was used for charge density, and "Physical Lines," 496, where the symbol e was used for charge density: "Now if e be the quantity of free electricity in unit of volume, then the equation of continuity will be. . . ."

43   See Maxwell, "Faraday's Lines," 195; see also Chapter 4 herein.

44   Maxwell to Thomson, 10 December 1861, in Larmor, ed., "Familiar Letters," 728–9: "I made out the equations in the country," i.e., at the Maxwell estate in "Galloway [southwestern Scotland] last summer [where] we spent all our three months vacation." In Maxwell to Faraday, 19 October 1861 – in Campbell and Garnett, *Life*, xxii – Maxwell similarly observed that he "worked out the formulae in the country." Cf. Everitt, *Maxwell*, 37–41, 98–9; Harman, "Consistent Representation," 191; Bromberg, "Displacement Current," 227; idem, "Maxwell's Concept of Electric Displacement," dissertation, University of Wisconsin, 1967 (University Microfilms No. 67-475), 176–8; see further Chapter 2, Section 3, herein.

45   Maxwell, "Physical Lines," 489–90.

46   Ibid., 489.

47   Ibid., 489–92.

48   See Boltzmann, ed., *Ueber physikalische Kraftlinien*, commentary on p. 126, where it is judged that if the elastic blobs are indeed allowed to bulge equatorially, the proper magnetic forces will be produced, with at most small changes in the numerical coefficients.

49   Maxwell, "Physical Lines," 490–6, quoted passage and final equation, equation (112), on 496; cf. equation (108) on 495.

50   Equation (3.14) from equations (113) and (114), ibid., 496–7. Equation (3.15a) from equation (115), ibid., 497; cf. equation (108) on 495; my ρ_m corresponds to Maxwell's e. Equation (3.15b) also from equation (115), ibid., 497, interpreted electromagnetically, my c corresponding to Maxwell's E. Schema (3.15) esp. from equation (118), ibid., 497; cf. equation (108) on 495.

51   Ibid., 497–8. I continue to transcribe Maxwell's E as c, to avoid confusion with the electric field; U, m, F, e_1, e_2, η_1, and η_2 are Maxwell's symbols. The relationship $\pi m = c^2$ used in equation (3.19) and throughout is from Maxwell's equation (108), on p. 495, our correspondence (3.15f). Note that in equation (3.17) – Maxwell's equation (116) – the energy U is positive, Maxwell's usual sign convention for P, Q, R (i.e., E_2 or τ_2) being employed, so that τ_2 and δ are oppositely directed – see Maxwell's equation (119). There is a sign error in Maxwell's equation (123), which is, however, without consequence, as it is not propagated to equation (124).

Our equations (3.16) and (3.19) correspond to Maxwell's equations (128) and (127), respectively, where the latter use the convention that F is the force that acts "against electric forces" in an electrostatic equilibrium situation, and thus are written with minus signs; both Maxwell's equations and ours signify, for charges of positive sign, "a repulsion varying inversely as the square of the distance." On units, see further Chapter 5.

52 Maxwell, *Treatise, 1*, 132 (1st ed.), 166 (3rd ed.): "I have not been able . . . to account by mechanical considerations for these stresses in the dielectric"; cf. 165 (3rd ed.), editor's note (the 3rd ed. was edited by J. J. Thomson).

4. The introduction of the displacement current

1 In "Physical Lines," 491, Maxwell referred to "the electric current due to displacement" – a concept to be explored later; the abbreviation to "displacement current" appears to be post-Maxwell.

2 Wise, "British Dynamical Theory." Representative of work in the past decades on the origin of the displacement current are Alfred M. Bork, "Maxwell, Displacement Current, and Symmetry," *American Journal of Physics, 31* (1963), 854–9; Bromberg, "Displacement Current"; Chalmers, "Maxwell's Methodology"; Hesse, "Logic of Discovery." Other contributions, including discussions of the later development of the displacement current, are listed in Wise's review.

3 Wise, "British Dynamical Theory," 190, 196.

4 See Chapter 1, Section 1, and Chapter 3.

5 Wise, "British Dynamical Theory," 196; Bromberg, "Displacement Current," 233.

6 Order and naming of the equations as in Jackson, *Classical Electrodynamics*, 177, see also 613–20. Cf. Wolfgang K. H. Panofsky and Melba Phillips, *Classical Electricity and Magnetism*, 2nd ed. (Reading, Mass.: Addison-Wesley, 1962), 160; Richard P. Feynman, Robert B. Leighton, and Matthew Sands, *The Feynman Lectures on Physics*, 3 vols. (Reading, Mass.: Addison-Wesley, 1964), *2*, 18-2, see also 32-3–32-5; and Edson Ruther Peck, *Electricity and Magnetism* (New York: McGraw-Hill, 1953), 429. [Equation (4.3), with $-\partial \mathbf{B}/\partial t$ on the right, as in Panofsky and Phillips, Feynman et al., and Peck.]

7 Panofsky and Phillips, *Electricity and Magnetism*, 28, 129, 135–6; Peck, *Electricity and Magnetism*, 68–72, 320–9; cf. Feynman et al., *Lectures, 2*, 32-3–32-5.

8 Jackson, *Classical Electrodynamics*, 133; Panofsky and Phillips, *Electricity and Magnetism*, 118; Peck, *Electricity and Magnetism*, 141–2; cf. Feynman et al., *Lectures, 2*, 18-2, 32-3–32-5.

9 Panofsky and Phillips, *Electricity and Magnetism*, 135–6; cf. Jackson, *Classical Electrodynamics*, 178, and Feynman et al., *Lectures, 2*, 32-3–32-5.

10 Jackson, *Classical Electrodynamics*, 178; Panofsky and Phillips, *Electricity and Magnetism*, 135, 136, 139, 140; Feynman et al., *Lectures, 2*, 18-1–18-3; Peck, *Electricity and Magnetism*, 426–7. As pointed out by Panofsky and Phillips, p. 139, one can imagine other ways of modifying Ampère's law for the nonstationary case; cf. Buchwald, *Maxwell to Microphysics*, 9; this point is, however, elided in most versions of the standard account.

11 See Panofsky and Phillips, *Electricity and Magnetism*, 136.

12 The macroscopic form of the equations is presented in Jackson, *Classical Electrodynamics*, 178, Panofsky and Phillips, *Electricity and Magnetism*, 160, and

Peck, *Electricity and Magnetism*, 429; Feynman et al., *Lectures*, 2, 32-3–32-5, regard them as outdated. Polarization currents as a part of the story are mentioned by Peck, pp. 427–8, and Panofsky and Phillips, pp. 135–6. More explicit discussion of polarization current as a motivation for displacement current occurs typically in lecture presentations and in the historical literature – e.g., Bromberg, "Displacement Current"; Chalmers, "Maxwell's Methodology"; and Hesse, "Logic of Discovery."

13 Maxwell, "Faraday's Lines," 190, 192, 194.

14 Maxwell used a left-handed coordinate system in "Faraday's Lines" – as explicitly described on p. 194 – and this has been taken into account in transcribing the equation. Also, the derivatives d/dx, d/dy, d/dz are clearly to be interpreted as partial derivatives, although no special notation was used to indicate this.

15 Ibid., 195.

16 Ibid.: "Our investigations are therefore for the present limited to closed currents; *and we know little of the magnetic effects of any currents which are not closed* [emphasis mine]."

17 Ibid., 155.

18 Ibid., 155–9, 188–9; Siegel, "Completeness as a Goal."

19 Maxwell, "Physical Lines," 451.

20 Ibid., 502, where the presentation of the equation is recapitulatory in nature, and it is presented in complete and canonical form. Cf. working forms of the equation for specific purposes on pp. 462 and 471; the former has the left- and right-hand sides reversed, but according to the accompanying text, as well as the placement of the $1/4\pi$ factor, maintains the field-primacy perspective.

21 Ibid., 496.

22 See Bromberg, "Displacement Current," esp. 220–3.

23 Maxwell, "Physical Lines," 496–7, equations (113)–(117).

24 Bromberg, "Displacement Current," esp. 222, recognizes the force of this argument, but still wants to see the solenoidal current of the standard account appearing at this point, at least in embryo. She is therefore forced to conclude that "the meaning of (p, q, r) is ambiguous."

25 Maxwell, "Physical Lines," 496–7.

26 Ibid., 498–9; see further Chapter 5, Section 3, esp. equation (5.8) and note 16.

27 Ibid., 496–8; cf. Bromberg, "Displacement Current," 223. Concerning Maxwell's hastening on without comment, cf., Ludwig Boltzmann's comments on Maxwell's gas theory, as quoted in E. Mendoza, ed., *A Random Walk in Science: An Anthology*, comp. R. L. Weber (London: Institute of Physics; New York: Crane, Russak, 1973), 43: "There is no time to say why this or why that substitution was made; he who cannot sense this should lay the book aside, for Maxwell is no writer of programme music obliged to set the explanation over the score." Concerning Maxwell's emphasis on embodied rather than disembodied mathematics, see Chapter 1 herein.

28 Maxwell, *Treatise*, 2, 232–3 (1st ed.), 253 (3rd ed.), where he refers to what we have called the solenoidal composite current as the "true electric current"; see further Chapter 6, Section 1, herein.

29 A comprehensive survey of these practices in the writing of equations would be beyond the scope of this study. Examples of Maxwell's own practices at the time,

from an area other than electromagnetic theory, are provided by his "Illustrations of the Dynamical Theory of Gases": In the equation $p = (\frac{1}{3})NMv^2$ (on p. 389), the gas pressure (p), as outcome or effect, stands alone on the left-hand side of the equation, while the parameters pertaining to the molecular impacts that cause or give rise that pressure – characterizing the masses (M), numbers (N), and velocities (v) of the particles – are combined on the right-hand side. In the equation $\mu = (1/3\sqrt{2})(Mv/\pi s^2)$ (on p. 391), gas viscosity (μ), as outcome or effect, stands alone on the left-hand side of the equation, while the parameters pertaining to the molecular momentum transport that gives rise to that viscosity – characterizing the masses (M), velocities (v), and sizes (s) of the particles – are combined on the right-hand side. Equations for diffusion (on p. 394) and thermal conductivity (on p. 404) display a similar pattern.

30 See, e.g., Panofsky and Phillips, *Electricity and Magnetism*, 136, "the magnetic effects of currents," and 329, "the source equations."

31 See, e.g., Tetu Hirosige, "Origins of Lorentz's Theory of Electrons and the Concept of the Electromagnetic Field," *Historical Studies in the Physical and Biological Sciences, 1* (1969), 151–209. New perspective is furnished by Buchwald, *Maxwell to Microphysics.*

32 See Chapter 1, Section 1; P. M. Heimann [Harman], "Maxwell, Hertz, and the Nature of Electricity," *Isis, 62* (1971), 149–57; Hesse, "Logic of Discovery"; A. F. Chalmers, "The Limitations of Maxwell's Electromagnetic Theory," *Isis, 64* (1973), 469–83. Cf. Chapter 6, Section 1.

33 Maxwell, "Faraday's Lines," 195.

34 Maxwell, "Physical Lines," 496.

35 Ibid., 490–1; Whittaker, *Aether and Electricity, 1,* 62–6; O. F. Mossotti, "Discussione analitica sull'influenza che l'azione di un mezzo dielettrico ha sulla distribuzione dell'elettricità alla superficie di più corpi elettrici disseminati in esso," *Società Italiana, Memorie, 24* (1847), 49–74; Faraday, *Experimental Researches, 2,* 360–416 (Series XI).

36 Maxwell, "Physical Lines," 491. Cf., for adumbration of the notion of electric currents existing in dielectric media, idem, "Faraday's Lines," 181, and Maxwell to Thomson, 13 November 1854, in Larmor, ed., "Familiar Letters," 701–5, esp. 704, also in Harman, ed., *Letters and Papers of Maxwell, 1,* 254–63, esp. 262.

37 Maxwell, "Physical Lines," 496.

38 Faraday, *Experimental Researches, 1,* 410.

39 See Chapter 1, Section 1; also see Harman, "Nature of Electricity"; Hesse, "Logic of Discovery"; cf. Peck, *Electricity and Magnetism,* 68–72.

40 See further Chapter 3, Section 2.

41 See especially Maxwell, "Physical Lines," 460–3 (Figures 3, 5, 6), 471–2, 477, and Plate VIII, opposite 488 (Figures 1, 2).

42 Peck, *Electricity and Magnetism,* 214–17, 244 (problem 7.55).

43 See Maxwell, "Physical Lines," 460 (Figure 3), 471–2, and Plate VIII, opposite 488 (Figure 1), for the circular cross sections and the three-dimensional relationships; also see p. 477, and Plate VIII, opposite 488 (Figure 2), for the successive rows of vortices in the plane, and the opposite senses of rotation on opposite sides. The variation of angular velocity with distance from the symmetry axis is described in Campbell and Garnett, *Life,* 534–5. Garnett, who wrote the account

of Maxwell's scientific achievements, had been Maxwell's first demonstrator at the Cavendish Laboratory – see pp. ix–x (editor's preface) – and thus presumably provides a knowledgeable contemporary account of the details of the model. Cf. Boltzmann, ed., *Ueber physikalische Kraftlinien*, 133. (Note that outside of the wire, the speed and direction of the vortex rotations vary in such a manner that the net flux of idle-wheel particles, averaged over pseudodifferential volumes, is zero; see further Appendix 2.)

44 Maxwell, "Physical Lines," 489–96. Various commentators read the text in various ways. Thus, Bromberg, "Displacement Current," 224, on the one hand, suggests that the pseudospherical elastic blobs of Part III undergo elastic distortion but do not rotate, whereas Boltzmann, ed., *Ueber physikalische Kraftlinien*, 133, has them both rotating and undergoing elastic distortion. The following circumstances lead me to side with Boltzmann: The term "molecular vortices" appears in the subtitle of Part III – Maxwell, "Physical Lines," 489 – implying that vortical or rotational motions persist there; the symbols α, β, γ, defined in Part I (pp. 457–8) as the components of a vector proportional to the angular velocities of the vortices, continue to be used in Part III (p. 496) – and indeed throughout all four parts of the paper – with no suggestion that their basic meaning designating rotations has changed; and the new term in Ampère's law – representing the effect of elastic distortions – is inserted in Part III (p. 496) as a "correct[ion]" term, *alongside* the **curl** term – representing the effect of rotations – rather than replacing it.

45 Maxwell, "Physical Lines," 496–7, 500–2; see also notes 41 and 43 in this chapter.

46 I follow here the line of analysis set out in Boltzmann, ed., *Ueber physikalische Kraftlinien*, 132–3. I do not, however, rely on Boltzmann's authority to support my own interpretation and elaboration of this line of analysis, but rather refer directly to the relevant primary source material, as cited in the preceding note 45 and throughout. See also Maxwell, "Physical Lines," 485–7, where motion of the small particles within the confines of a material molecule is envisioned; in Part III, however, complete immobilization of the small particles in a dielectric medium is required, in order to allow for complete identification of static electric charge with polarization-bound charge in the magnetoelectric medium – see Section 6, this chapter.

47 Cf. Boltzmann, ed., *Ueber physikalische Kraftlinien*, 133.

48 Maxwell, "Physical Lines," 474–84, 490.

49 Ibid., 496–7, equations (112), (113), and (115). The equations in **E** are the most direct transcriptions; the equations in **P** are obtained by using $\mathbf{E} = -4\pi c^2 \mathbf{P}$ (Section 4, this chapter).

50 Ibid., 491–5.

51 Ibid., 496; see also the preceding note 44.

52 Ibid., 491.

53 Ibid., 492.

54 Ibid., 494–5.

55 Ibid., 495, equations (99)–(103).

56 Ibid., 495, equations (104)–(106).

57 Ibid., 493–5. Maxwell to Faraday, 19 October 1861, in Campbell and Garnett, *Life*, xx–xxii.

58 Maxwell, "Physical Lines," 497–502; see also Chapter 3, Section 4.
59 Cf. Bromberg, "Maxwell's Electrostatics."

5. The origin of the electromagnetic theory of light

1 Geoffrey Cantor, *Optics after Newton: Theories of Light in Britain and Ireland, 1704–1840* (Manchester University Press, 1983), 78, 104, 112, 123–4, 128, 134, and passim. Chapter 1, Section 1 herein. Williams, *Faraday,* 148–51.
2 Chapter 2, Section 2; Michael Faraday, "Thoughts on Ray-Vibrations" (1846), *Experimental Researches, 3,* 447–52; Siegel, "Universal Ether," 242–6. Cf. also William John Macquorn Rankine, "On the Vibrations of Plane-Polarized Light" (1851), in *Scientific Papers,* 150–5, for a proposed connection of molecular vortices with light, if not with electricity and magnetism.
3 Christa Jungnickel and Russell McCormmach, *Intellectual Mastery of Nature: Theoretical Physics from Ohm to Einstein,* 2 vols. (University of Chicago Press, 1986) *1, The Torch of Mathematics, 1800–1870,* 142–6; Rosenfeld, "Evolution of Electrodynamics," 1633–4. (I follow Rosenfeld in designating Weber's constant c as c_W, in order to distinguish it from the modern c; for Maxwell's E, see Chapter 4, Section 2, herein; for his v, see Maxwell, "Dynamical Theory," 569.) A preliminary value, $c_W = 436{,}090 \times 10^6$ millimeters/second, was presented in Wilhelm Weber, "Vorwort bei der Uebergabe der Abhandlung: Elektrodynamische Maassbestimmungen, insbesondere Zurückführung der Stromintensitäts-Messungen auf mechanisches Maass" (1855), in *Wilhelm Weber's Werke, 3,* ed. Heinrich Weber (Berlin: Julius Springer, 1893), 591–6, esp. 594. The definitive value, $c_W = 439{,}450 \times 10^6$ millimeters/second, was presented in Wilhelm Weber and Rudolph Kohlrausch, "Ueber die Elektricitätsmenge, welche bei galvanischem Strömen durch den Querschnitt der Kette fliesst" (1856), in Weber, *Werke, 3,* 597–608, on 605, and also in Rudolph Kohlrausch and Wilhelm Weber, "Elektrodynamische Maassbestimmungen insbesondere Zurückführung der Stromintensitäts-Messungen auf mechanisches Maass," (1857), in Weber, *Werke, 3,* 609–76, on 652.
4 Gustav Kirchhoff, "Ueber die Bewegung der Elektricität in Drähten" (1857), and "Ueber die Bewegung der Elektricität in Leitern" (1857), in *Gesammelte Abhandlungen* (Leipzig: Johann Ambrosius Barth, 1882), 131–54, 154–68, on 142–3, 146–7. The conversion of the German miles per second into meters per second is taken from a later paper by Wilhelm Weber, "Elektrodynamische Maassbestimmungen insbesondere über elektrische Schwingungen" (1864), in *Wilhelm Weber's Werke, 4,* ed. Heinrich Weber (Berlin: Julius Springer, 1894), 104–241, on 157, where c_W is given as $439{,}450 \times 10^6$ millimeters/second, and $c_W/\sqrt{2}$ as $310{,}740 \times 10^6$ millimeters/second or 41,950 German miles per second. (It follows that Kirchhoff must have used more significant figures than he quoted for c_W in his calculation of the velocity as 41,950 German miles per second.) Jungnickel and McCormmach, *Intellectual Mastery, 1,* 296–7; Rosenfeld, "Evolution of Electrodynamics," 1634–41.
5 Jungnickel and McCormmach, *Intellectual Mastery, 1,* 296–7 (quotation of Kirchhoff on p. 297). Rosenfeld, "Evolution of Electrodynamics," 1634–41 ("extraordinary failure" on p. 1635). Leon Rosenfeld (paper delivered at the International Symposium on Nineteenth-Century Physics, Aarhus, Denmark, August 1970). In 1864 – somewhat after the fact for our purposes – Weber published a

calculation of "electrical oscillations" paralleling Kirchhoff's, and arriving at exactly the same numerical result: Weber, "Elektrodynamische Maassbestimmungen," (1864), on 157. Weber's comment, on pp. 157–8, is revealing: "If this close agreement between the velocity of propagation of electric waves and that of light could be regarded as an indication of an inner connection of the two subjects, it would command the greatest interest, because of the great importance that the investigation of such a connection would have. It is evident, however, that in this regard, above all, the true significance of that velocity in its connection with electricity must be taken into account, and this does not appear to be such as to arouse great expectations. For the approximation of the true velocity of propagation to the limiting value that agrees with the velocity of light requires, as was shown, not only that the conducting wire be very thin in comparison with its length, but also that this long and thin wire have very small resistance. It becomes evident from this that a close approximation to that limiting value will seldom obtain, while large deviations from it will be very common."

6 It is not clear precisely when Maxwell became aware of Kirchhoff's paper of 1857. Maxwell did, in 1868, cite a paper by Ludwig Lorenz, published in 1867, which stemmed from Kirchhoff's work – Maxwell, "Direct Comparison; Note on the Electromagnetic Theory of Light," 137 – and Kirchhoff is directly cited in Maxwell, *Treatise*, 2, 398 (1st ed.), 450 (3rd ed.). See further Chapter 6, Section 3, herein.

7 Representative of the spectrum are Duhem, *Aim and Structure*, 98, and idem, *Théories Electriques*, at one extreme – seeing the model as having little substantive role; then Chalmers, "Maxwell's Methodology," esp. 137, where the model is seen as having some slight utility; Bromberg, "Displacement Current," where the model is seen as having greater heuristic value, although in a primarily accommodative mode; and, tending more toward seeing the model as determinative, Boltzmann, ed., *Ueber physikalische Kraftlinien*, and the present work.

8 Equations explicit, e.g., in Whittaker, *Aether and Electricity, 1*, 252, and M. S. Longair, *Theoretical Concepts in Physics: An Alternative View of Theoretical Reasoning in Physics for Final-Year Undergraduates* (Cambridge University Press, 1984), 47–8. In Tolstoy, *Maxwell*, 124, the "undulating magnetic and electric fields transverse to their direction of propagation" are explicit, as is the characterization of the mechanical model as a bow to nineteenth-century style: "The thought habits of his [Maxwell's] era demanded some sort of mechanism, however outlandish – everyone was offering up aether models; it was, one might say, the accepted language of the day."

9 Bromberg, "Displacement Current," 219.

10 Panofsky and Phillips, *Electricity and Magnetism*, 185–6; the argument is carried through for **E** and **B** in Peck, *Electricity and Magnetism*, 429–36, and Jackson, *Classical Electrodynamics*, 202–5. Feynman et al., *Lectures*, 2, 32-3–32-5, suggest that making use of **H** is the traditional, Maxwellian choice, while using **B** is the more appropriate modern choice.

11 Maxwell, "Physical Lines," 499, 509.

12 See Maxwell's averaging over pseudodifferential volumes containing many vortices, ibid., 469–71.

13 Bromberg, "Displacement Current," 227, finds the model accommodative in a

somewhat different sense: Supplying a second set of variables – the mechanical variables – to accompany the electromagnetic variables, the model allowed for "double vision" on Maxwell's part, thereby accommodating a kind of fruitful confusion concerning the signs in the equations, so that, for example, the vector (P, Q, R) was "simultaneously an electric force in the direction of the displacement and an elastic force opposite to it"; cf. Chapter 4.

14   Maxwell, "Physical Lines," 495, 499, equations (133), (108), (132), (136).

15   Maxwell's definition of the ratio of units, ibid., 498–9, defined unit current in electromagnetic units in terms of a current loop of unit area equivalent to a unit magnetic dipole; thus, here and throughout – see, e.g., *Treatise*, 2, 3, 239 (1st ed.), 3, 263 (3rd ed.) – the definition of the electromagnetic unit was based ultimately on magnetic poles. Maxwell's definition is, however, equivalent to a modern definition in terms of forces between parallel currents, given the equivalence of a current loop of unit strength and unit area to a unit dipole – e.g., Jackson, *Classical Electrodynamics*, 613–14, 132–7, 141–3.

16   Maxwell, "Physical Lines," 499; as in Chapters 3 and 4, I have for mnemonic purposes transcribed Maxwell's E as c. Maxwell took the data from Kohlrausch and Weber, "Elektrodynamische Maassbestimmungen" (1857), 651, where they evaluated a parameter equal to $c_w/2\sqrt{2}$ as $155,370 \times 10^6$ millimeters per second; Maxwell cited this parameter and multiplied it by two. Although the identification of Maxwell's E as the modern c is clear enough on physical and mathematical grounds, there is some semantic confusion: Maxwell referred to E as "the number by which the electrodynamic measure of any quantity of electricity must be multiplied to obtain its electrostatic measure," which invites confounding it with Weber's c_w, which, in Weber's terminology, measures the ratio of "electrodynamic" and "electrostatic" units (pp. 651–2); this is, however, only a problem in terminology, which was not uniform at that time; in later work, Maxwell referred to "310,740,000,000 metres per second" as "the number of electrostatic units in one electromagnetic unit," thus distinguishing it terminologically from Weber's c_w – Maxwell, "Dynamical Theory," 569.

17   Jackson, *Classical Electrodynamics*, 614.

18   Maxwell to Thomson, 10 December 1861, in Larmor, ed., "Familiar Letters," 729.

19   This is clearest in Maxwell's equations (101) and (102), "Physical Lines," 495–7, where the displacement δ is proportional jointly to the surface density of the small particles σ and the strain parameter t, and equation (115) taken together with equation (119), which makes ρ_P linear in δ.

20   For the mechanical calculation of torsion waves in a uniform elastic medium, see, e.g., Arnold Sommerfeld, *Lectures on Theoretical Physics. Vol. 2: Mechanics of Deformable Bodies*, tr. G. Kuerti (New York: Academic Press, 1950), 105–7. The definition of the shear or torsion modulus – Maxwell's m – to be used in this equation is a matter of some delicacy, easily subject to an error of a factor of 2 – see pp. 8–9, 67–9, the issue of the angle γ being "*twice* the strain component" ϵ_{xy} (emphasis Sommerfeld's). Note also the rampant factors of 2 in the table on p. 9 comparing different systems of notation. It will become evident that Maxwell did in fact make an error of a factor of 2 in this connection.

21   Maxwell, "Physical Lines," 500, where the quoted value for Fizeau's measure-

ment of the velocity of light, "70,843 leagues per second (25 leagues to a degree)," is referred to "*Comptes Rendus,* Vol. xxix (1849), p. 90." However, in H. Fizeau, "Sur une expérience relative à la vitesse de propagation de la lumière," *Comptes Rendus Hebdomadaires des Séances de L'Académie des Sciences, 29* (1849), 90–2, on 92, the result is stated as "70948 lieues de 25 au degré." The garbling of the lower-order digits in Maxwell's report of the result does not materially affect his argument, for the difference between 70,843 and 70,948 is of the order of 0.1%, while the agreement between Fizeau's number and the Weber-Kohlrausch number is only at the order of 1.0% (as discussed later); this does give some feeling, however, for the kinds of numerical errors that Maxwell made, which will be important in the sequel.

22　Maxwell also cited, in a note – "Physical Lines," 500n – another reported value of Fizeau's result for the velocity of light, 193,118 miles per second, which differs from the calculated velocity of magnetoelectric undulations by only 0.02%; Maxwell quoted also the velocity of light as "deduced from aberration," namely, 192,000 miles per second, which differs by 0.6%.

23　Ibid., 469–70, equation (27), characterization of surface element dS as "separating the first and second vortices," and corresponding treatment of direction cosines in equation (29).

24　If the medium is space-filling, as in Figure 2.2, and we are considering a region of zero conductivity, where the small spherical particles are pinned in place, and if, further, there are no electric or magnetic fields in the region, so that there are no relative motions or deformations at the level of individual cells, then the cellular structure of the medium will become moot, and the medium will evidently behave as an unarticulated, uniform elastic solid, sustaining torsion waves as in Figure 5.1. Cf. the discussion of torsion waves, corresponding to light but *not* involving electric and magnetic fields, in Section 2 of this chapter.

25　See Chapter 3, Section 1, especially the discussion relating to equation (3.6); cf. Appendix 1. The specialization in both places to uniform density and angular velocity was supported by the explicit specification of a homogeneous elastic solid material (with nonvarying elastic constants) for the contents of the vortex cells in Part III of Maxwell, "Physical Lines," 493–6; the transition from circular cylinders to spheres was left as an approximation.

26　Ibid., 492. Pertaining to linkage (L1′), Boltzmann, ed., *Ueber physikalische Kraftlinien,* 90–2, estimates errors, in the approximation of a system of space-filling cylindrical vortices (of square or hexagonal cross section) to circular ones, amounting to $4/\pi = 1.27$ for the square array and $2\sqrt{3}/\pi = 1.10$ for the hexagonal array – with the transition to spheres to be further compounded. The volume approximation enters into (L2′) – Maxwell, "Physical Lines," 470. Both surface and volume approximations enter into (L3′) and (L4′) – pp. 495–7. And (L5′), which depends on the elastic-energy calculation, goes by way of either the volume approximation or the surface approximation – pp. 493–4. Also providing some sense for the numerical factors involved in alternative calculations concerning the vortices is the MS "Prop XII" (see Appendix 1), where the result for the ratio of the angular momentum of a vortex to its energy can vary by a factor $2/1.178 = 1.7$.

27　Duhem, *Théories Electriques,* 208–12. According to Maxwell, "The Equilibrium of Elastic Solids" (1850), in Niven, ed., *Scientific Papers of Maxwell, 1,* 30–73, esp. 44–5, the recommended method for the measurement of *m* is by measurements on the torsion of a circular cylinder; comparing Maxwell's equations here with the analogous equations in, e.g., Sommerfeld, *Mechanics of Deformable Bodies,* 302–3, 67–8, we find that $m = 2\mu$, where μ is Sommerfeld's "torsion modulus," or "shear modulus," or "Lamé's modulus," such that the velocity of torsion waves is given by $\sqrt{\mu/\rho}$; Maxwell thus should have used $m/2$ rather than *m* in equation (5.7), which is exactly Duhem's point. (See also note 20.)

28 Maxwell, "Physical Lines," 495–6; cf. Sommerfeld, *Mechanics of Deformable Bodies,* 61–9. Aside from the adjustability of the parameter *m,* one must point to another aspect of theoretical calculations in general: Just as an experimentalist will sometimes continue to take data until he has a nice answer, and then stop – Robert Olby, *Origins of Mendelism,* 2nd ed. (University of Chicago Press, 1985), 209–11 – so also will a theoretician sometimes continue to calculate – that is, to make refinements and hunt errors – until he has a nice answer, and then stop. One may guess that Maxwell did something like that; in particular, his stopping before he found the error in the definition of *m* might have had something to do with his having already gotten a neat answer, and therefore not being motivated to consider the matter further. In this sense, I would agree with Duhem that Maxwell was drawn on by the nice answer that he wanted to get; I do not think, however, that one need conclude that Maxwell consciously "went so far as to falsify one of the fundamental formulas of elasticity" – Duhem, *Aim and Structure,* 98 – just as one need not be pushed toward the conclusion that Gregor Mendel's too-neat data were "falsified" by "some assistant" or other parties unspecified – R. A. Fisher, "Has Mendel's Work Been Rediscovered?" *Annals of Science, 1* (1936), 115–37, on 132.

29 Maxwell to Faraday, 19 October 1861, in Campbell and Garnett, *Life,* xxii. Most commentators believe Maxwell's statements to the effect that he had not anticipated the close agreement between the ratio of units and the velocity of light – see, e.g., Bromberg, "Displacement Current," 227; Harman, "Modes of Consistent Representation," 192n; Everitt, *Maxwell,* 99. Chalmers, "Maxwell's Methodology," 136n, is more cautious and agnostic. The modern historian of science wants to avoid hagiography, but in making a judgment on Maxwell's veracity in this matter, one does want to take into account his strong religious conviction and earnest character, his dedication to science and to duty in general, his great respect for Faraday and Thomson, to whom he made these statements, and other such circumstances, as detailed in Campbell and Garnett, *Life,* passim; Everitt, *Maxwell,* passim; idem, "Maxwell's Creativity," passim.

On the importance of the element of surprise, as discussed by Herschel, Whewell, and Mill – all of whom Maxwell had read (see Chapter 1) – see Laudan, "The Medium and Its Message," 173–80.

30 Maxwell to Faraday, 19 October 1861, in Campbell and Garnett, *Life,* xxii (emphasis mine).

31 Maxwell to Thomson, 10 December 1861, in Larmor, ed., "Familiar Letters," 729; the "Weber's number" referred to here is the parameter equal to $c_w/2\sqrt{2}$ cited in note 19.

32 Fizeau, "Sur une expérience," 92.

33 Maxwell, "Physical Lines," 500–1.
34 Herschel, *Preliminary Discourse*, 29–34; Whewell, *Philosophy of the Inductive Sciences*, 62–8; Wilson, "Reception of the Wave Theory," 51; Laudan, "The Medium and Its Message," 173–80.
35 Maxwell to Faraday, 19 October 1861, in Campbell and Garnett, *Life*, xx–xxi; Maxwell to Thomson, 10 December 1861 in Larmor, ed., "Familiar Letters," 729 (see also Larmor's note); Maxwell to Monro, 18 February 1862, in Campbell and Garnett, *Life*, 334.
36 Maxwell, "Physical Lines," 506–13; Knudsen, "Faraday Effect," 255–61.

6. Beyond molecular vortices

1 See further Chapter 2, Section 5, and the Conclusion to this volume.
2 See Chapter 4, Section 2.
3 Maxwell to Droop, 28 December 1861, also 24 January 1862, in Campbell and Garnett, *Life*, 330–1 (emphasis mine), also 291, 332, 342 (other letters to Droop); Everitt, *Maxwell*, 79; J. A. Venn, *Alumni Cantabrigiensis: A Biographical List of All Known Students, Graduates and Holders of Office at the University of Cambridge, from the Earliest Times to 1900, Part II, from 1752 to 1900*, 4 vols. (Cambridge University Press, 1940–51), s.v. "Droop, Henry Richmond."
4 Maxwell, "Dynamical Theory," and "Direct Comparison; Note on the Electromagnetic Theory of Light."
5 The documents characterized in Appendix 2 are, however, of some relevance and will be cited.
6 Maxwell, "Dynamical Theory," 560; in Maxwell's system of units here, as in modern electromagnetic units, a factor of c^2 intervenes between $\mathbf{D}_M$ or $\mathbf{D}$ and $\mathbf{E}$, and is included in ϵ or k – compare ibid., 569, and Jackson, *Classical Electrodynamics*, 617–18. The change to the positive sign in equations (6.1) may be traced in the manuscript material (see Appendix 2): The sign is negative in "[DT, B]," but positive in "[DT, A]" and "[DT, D]"; the change in notation for the dielectric constant and the ratio of units may be traced through "[DT, C]." In "Dynamical Theory [MS]," equations (6.1) are written with an explicit plus sign, emphasizing the change; and the old "$4\pi E^2$" notation for dielectric constant and ratio of units appears but is crossed out (this is at the portion of "Dynamical Theory [MS]" corresponding to "Dynamical Theory," 570).
7 Maxwell, "Dynamical Theory," 554, 561, and "Direct Comparison; Note on the Electromagnetic Theory of Light," 139. The rationale for the displacement current in "Dynamical Theory" – as emancipated from the details of the molecular-vortex model – is summarized succinctly in Everitt, "Maxwell's Creativity," 128–9.
8 Maxwell, "Dynamical Theory," 557, also 578, where equations (19) and (20) – in component form – are "combine[d]."
9 Ibid., 561, equation (G).
10 Ibid., 554, 557, 561–2. It was argued earlier, especially in Chapter 4, that a full analysis of "Physical Lines" shows that the inconsistencies in algebraic signs appearing in that paper were not fundamental, and can and should be explained away. The inconsistencies in "Dynamical Theory," however, would appear to be more recalcitrant: Here it is Maxwell's final set of equations – pulled out in a separate list and distinguished from the other equations by letter designations

rather than numbers – that is internally inconsistent; this is to be distinguished from "Physical Lines," where the inconsistencies are localized in one set of intermediate steps, and the final equations are completely consistent, although different from the modern equations. I have not, however, undertaken the kind of exhaustive analysis of "Dynamical Theory" that I have of "Physical Lines," and it could be that such an analysis would disclose more consistency in "Dynamical Theory" than I have seen. My judgment of inconsistency must therefore be taken as provisional, although not unfounded. For parallel conclusions on the inconsistency in "Dynamical Theory," see, e.g., Chalmers, "Maxwell's Methodology," 141–2; Bromberg, "Maxwell's Electrostatics," 145–8; Hendry, *Maxwell*, 212–13.

11 Cf. Hendry, *Maxwell*, 210–15; Hesse, "Logic of Discovery," 102–5. The dilemma spilled over also into the equation relating electric field and conduction current, which Maxwell wrote with a positive sign in the manuscript "[DT, A]," but with a negative sign in the published "Dynamical Theory," 561. The sense of dilemma is conveyed in the handwriting in "[DT, B]," where Maxwell wrote equations (6.1) with the old negative sign, compatible with his view of the nature of electric charge, but not with his reformulation of Ampère's law (see Appendix 2).

12 Maxwell, *Treatise, 2,* 227–38 (1st ed.), 247–59 (3rd ed.), equation (J) on 233 (1st ed.), 254 (3rd ed.),

$$\rho = \frac{df}{dx} + \frac{dg}{dy} + \frac{dh}{dz}$$

quaternion form on 238 (1st ed.), 259 (3rd ed.).

13 "Electricity" and "charge" as in Maxwell, *Treatise, 1,* 166 (3rd ed.); the earlier version, *1,* 132–3 (1st ed.), differs somewhat in terminology.

14 Ibid., *1,* 166–8 (3rd ed.); the earlier version, *1,* 132–4 (1st ed.), employs somewhat different terminology and is a bit less clear.

15 There seems to be general agreement among Maxwell scholars that "Maxwell's ideas of charge and displacement in the *Treatise* . . . are quite coherent," although different from their modern counterparts – Wise, "British Dynamical Theory," 196, relying on Buchwald and Hesse; Buchwald, *Maxwell to Microphysics,* esp. 23–34; Bromberg, "Concept of Electric Displacement," esp. 109–27; cf., however, idem, "Maxwell's Electrostatics."

16 Maxwell to Faraday, 19 October 1861, in Campbell and Garnett, *Life,* xx–xxii; Maxwell to Thomson, 10 December 1861, in Larmor, ed., "Familiar Letters," 728–30; Monro to Maxwell, 23 October 1861, in Campbell and Garnett, *Life,* 329; Maxwell to Hockin, 7 September 1864, in Campbell and Garnett, *Life,* 340; Maxwell, "Dynamical Theory," esp. 577–88.

Cecil James Monro received his Cambridge B.A. a year after Maxwell (38th wrangler, 1855); Monro, "on account of ill-health was forced to live largely abroad, but devoted himself to the work of letter-writing and criticism of proof-sheets for friends" – Venn, *Alumni Cantabrigiensis,* s.v. "Monro, Cecil James." See further Campbell and Garnett, *Life,* 201–2 and passim.

17 See Everitt, *Maxwell,* 99–101, on Maxwell's work in this area beginning in 1862, and the publication, with Fleeming Jenkin, of "On the Elementary Relations of Electrical Quantities" in 1863; subsequent publications on this subject include Maxwell, "Direct Comparison; Note on the Electromagnetic Theory of Light,"

125–36; idem, *Treatise, 2*, 239–45, 334–82 (1st ed.), 263–9, 374–430 (3rd ed.).

18   Maxwell, "Dynamical Theory," 577–9; cf. Panofsky and Phillips, *Electricity and Magnetism*, 240–1.

19   Maxwell, "Direct Comparison; Note on the Electromagnetic Theory of Light," 138, 140–1.

20   Maxwell, "Dynamical Theory," 579–80; Maxwell also gave Fizeau's number, as 314,858,000, and the aberration value, as 308,000,000, for the velocity of light. Although Maxwell here avoided any "hypothes[e]s" of a "particular" nature concerning the mechanical substratum of the field, he did still believe it was there – pp. 563–4; cf. 528–33. See also Chapter 2, Section 5.

21   Maxwell, *Treatise, 2*, 387 (1st ed.), 436 (3rd ed.).

22   Ibid., *2*, 387–8 (1st ed.), 436 (3rd ed.).

23   Ibid., *2*, 388–9 (1st ed.), 437–8 (3rd ed.).

24   Ibid., *2*, 399–417, on 415 (1st ed.), 451–70, on 468 (3rd ed.); Knudsen, "Faraday Effect," esp. 255–61; cf. also Harman, "Consistent Representation," 193–5, 202–3.

25   Hermann Helmholtz, "Ueber die Bewegungsgleichungen der Elektricität für ruhende leitende Körper" (1870), in *Wissenschaftliche Abhandlungen von Hermann Helmholtz*, 3 vols. (Leipzig: Johann Ambrosius Barth, 1882–95), *1*, 545–628, esp. 548, citing Maxwell, "Dynamical Theory." Buchwald, *Maxwell to Microphysics*, esp. 73, 177.

26   Romualdas Sviedrys, "The Rise of Physical Science at Victorian Cambridge [with commentary by Arnold Thackray and reply by Romualdas Sviedrys]," *Historical Studies in the Physical and Biological Sciences, 2* (1970), 127–52, esp. 141–2, 149–50; Buchwald, *Maxwell to Microphysics*, 73–74ff.; Campbell and Garnett, *Life*, 348–66; Wilson, "Physics in the Natural Sciences Tripos," 343–53.

27   Knudsen, "Thomson's Electromagnetic Theory," 167–8, 172–3. William Thomson, "A Simple Hypothesis for Electro-Magnetic Induction of Incomplete Circuits, with Consequent Equations of Electric Motion in Fixed Homogeneous Solid Matter" (1888), in *Mathematical and Physical Papers, 4*, 539–44, on 543–4. M. Norton Wise and Crosbie Smith, "The Practical Imperative: Kelvin Challenges the Maxwellians," in Kargon and Achinstein, eds., *Kelvin's Baltimore Lectures*, 323–48. (Pages 7–263 of the latter comprise the text of A. S. Hathaway's notes of the lectures given by Kelvin at Johns Hopkins in 1884 and will be referred to herein as Kelvin, *Baltimore Lectures [1884]*.)

28   Kelvin, *Baltimore Lectures [1884]*, on 12, also 41–2, but cf. 206; Siegel, "Universal Ether," 242, 244–6, 254–7; Knudsen, "Thomson's Electromagnetic Theory," 171–5; Wise and Smith, "Practical Imperative"; Harman, "Maxwell's Dynamical Physics," esp. 267–9, 290–1.

29   Kelvin to Fitzgerald, 9 April 1896, in Thompson, *Life of Kelvin, 2*, 1064–9, on 1065; Kelvin, *Baltimore Lectures [1884]*, on 206; Knudsen, "Thomson's Electromagnetic Theory," 176–8; Siegel, "Universal Ether," 262–3.

30   Perhaps the most piquant expression, in Maxwell's *Treatise*, of his opinion concerning the existence of primitive, microscopic charge carriers was in connection with his discussion of electrolysis, where, though he admitted that the notion of a "molecule of electricity" was of heuristic value, he cautioned the reader that "this phrase [was] gross . . . and out of harmony with the rest of this treatise" – *1*,

312 (1st ed.), 380 (3rd ed.). (See *Oxford English Dictionary*, s.v. "gross": "**1777** PRIESTLEY *Philos. Necess.* 189 You will blush when you reflect a moment upon things so very gross as these. **1791** BOSWELL *Johnson* Jan. an. 1749, Some of them [Juvenal's Satires] . . . were too gross for imitation.") Kelvin quoted this passage from Maxwell's *Treatise* as showing that "Maxwell's electromagnetic theory of light was essentially molar," and hence incapable of giving an appropriate, molecular account of the interaction of light and matter – Lord Kelvin, *Baltimore Lectures on Molecular Dynamics and the Wave Theory of Light* (London: C. J. Clay, 1904), 376–7. Knudsen, "Thomson's Electromagnetic Theory," 175–6. Buchwald, *Maxwell to Microphysics*, ix–xiiiff.

31 Chapter 1, Section 1; Williams, *Origins of Field Theory*, 96–115; Chapter 5, Section 1; Helmholtz, "Ueber die Bewegungsgleichungen der Elektricität."

32 Hermann Helmholtz, "Ueber die Erhaltung der Kraft: Eine physikalische Abhandlung" (1847), in *Wissenschaftliche Abhandlungen, 1*, 12–68, esp. 26–7, tr. John Tyndall as "The Conservation of Force," in Brush, ed., *Kinetic Theory, 1*, 89–110, esp. 102. A. E. Woodruff, "The Contributions of Hermann von Helmholtz to Electrodynamics," *Isis, 59* (1968), 300–10, esp. 304–5. Wise, "German Concepts," 287–99. Maxwell, "Faraday's Lines," 208; "Physical Lines," 488; "Direct Comparison; Note on the Electromagnetic Theory of Light," 137–8. Archibald, "Action-at-a-Distance Electromagnetic Theory," 365–416.

33 Rosenfeld, "Evolution of Electrodynamics," 1634–8; Jungnickel and McCormmach, *Intellectual Mastery, 1*, 174–85. Archibald, "Action-at-a-Distance Electromagnetic Theory," 312–36.

34 Wise, "German Concepts," 294–5; Archibald, "Action-at-a-Distance Electromagnetic Theory," 324–34; Helmholtz, "Ueber die Bewegungsgleichungen der Electricität," 548.

35 Wise, "German Concepts," 293–4, 297–8.

36 Helmholtz, "Ueber die Bewegungsgleichungen der Electricität," esp. 627–8. Buchwald, *Maxwell to Microphysics*, 177–86. Hertz, *Electric Waves*, 20–8, esp. 24–5.

37 Buchwald, *Maxwell to Microphysics*, 177–268. On the question of electromagnetic waves, cf. Chalmers, "Limitations of Maxwell's Electromagnetic Theory," esp. 471–6. Chalmers also believes that certain characteristic aspects of Maxwell's approach to electromagnetic theory prevented him from appreciating the possibility of producing electromagnetic waves, but Chalmers emphasizes "two factors: the strong sense in which [Maxwell] aimed to reduce all the physical sciences to mechanics and his reluctance to make hypotheses" – p. 475. Chalmers's analysis and my own intersect in the judgment that Maxwell would not have subscribed to a picture of light originating from the motions of microscopic charges in molecules. As Chalmers does point out, however, Maxwellian followers such as George Fitzgerald and Oliver Lodge ultimately did consider the possibility of generating electromagnetic waves – p. 476. See also Thomas K. Simpson, "Maxwell and the Direct Experimental Test of His Electromagnetic Theory," *Isis, 57* (1966), 411–32.

38 Recent and authoritative is Buchwald, *Maxwell to Microphysics;* see also Russell McCormmach, "H. A. Lorentz and the Electromagnetic View of Nature," *Isis, 61* (1970), 459–97. On the background to relativity theory, see, e.g., the classic arti-

cles by Gerald Holton in *Thematic Origins of Scientific Thought: Kepler to Einstein* (Cambridge, Mass.: Harvard University Press, 1973). Somewhat tangential, but reflecting my own views, and bibliographically relevant, is Daniel M. Siegel, "Classical-Electromagnetic and Relativistic Approaches to the Problem of Nonintegral Atomic Masses," *Historical Studies in the Physical and Biological Sciences, 9* (1978), 323–60.

39   Hertz, *Electric Waves*, 1–7, on 1, 6. Buchwald, *Maxwell to Microphysics*, 75; John David Miller, "Rowland and the Nature of Electric Currents," *Isis, 63* (1972), 5–27, esp. 10–12. Abraham Pais, *'Subtle is the Lord . . . ': The Science and the Life of Albert Einstein* (Oxford University Press, 1982), 112.

40   Hertz, *Electric Waves*, 1–14; idem, "On Electromagnetic Waves in Air and Their Reflection" (1888), in *Electric Waves*, 124–36.

41   Tetu Hirosige, "Origins of Lorentz' Theory of Electrons and the Concept of the Electromagnetic Field," 151–209; Buchwald, *Maxwell to Microphysics.* Hugh G. J. Aitken, *Syntony and Spark – The Origins of Radio,* (New York: Wiley, 1976).

42   See Chapter 4, esp. Section 1.

43   Hertz, *Electric Waves,* 21; Harman, *Energy, Force, Matter,* 109–11, 149–55. Russell McCormmach, "Einstein, Lorentz, and the Electron Theory," *Historical Studies in the Physical and Biological Sciences, 2* (1970), 41–87. Martin J. Klein, "Mechanical Explanation at the End of the Nineteenth Century," *Centaurus, 17* (1973), 58–82, esp. 69–70, 74–5, 78. Siegel, "Universal Ether," 262–3; McCormmach, "Electromagnetic View of Nature"; Stanley Goldberg, "In Defense of Ether: The British Response to Einstein's Special Theory of Relativity, 1905–1911," *Historical Studies in the Physical and Biological Sciences, 2* (1970), 89–125.

Conclusion

1   Hendry, *Maxwell,* 272, referring to Siegel, "Universal Ether," understands Maxwell to be there characterized as a "vacillating or inconsistent mechanist"; the intent, however, both in "Universal Ether" and herein (see esp. Chapter 2), has been to characterize Maxwell rather as a growing and developing mechanist.

2   Viewed from a retrospective standpoint, the aptness and utility of the molecular-vortex model might be characterized as follows: As concerns the electromagnetic theory of light, the aptness of the model can be seen in energetic terms. Maxwell's mechanical medium was capable of storing energy in two ways: as the kinetic energy of the whirling vortices, corresponding to magnetic field energy; and as the potential energy of the elastically deformed vortex material, corresponding to electric field energy. Regulating the trade-off between these two kinds of energy, in ordinary electromagnetic processes such as the charging of a capacitor, or in the propagation of undulatory waves in the medium, was the constant c, corresponding to the ratio of units. Thus stated in energetic terms, the situation in Maxwell's original mechanical theory of the magnetoelectric medium is quite similar to the situation envisioned in his later – and our modern – electromagnetic theory of light, where there are again two kinds of energy – electrical and magnetic field energy – which trade off against each other in the propagation of electromagnetic waves, as again regulated by the constant c. This energetic parallel-

ism between the original, mechanically mediated electromagnetic theory of light and the subsequent, purely electromagnetic theory accounts for the fact that the former could serve, in a manner parallel to the latter, to establish a relationship between the ratio of units and the velocity of light. As concerns the displacement current, energetic considerations and the aptness of the movable idle-wheel particles for modeling charge and current (this only from a modern point of view, not from a later Maxwellian point of view) combine to make the molecular-vortex model useful in modeling the electromagnetic connections of the displacement current.

3   O. J. Lodge, "[From the obituary notices of the Royal Society]," in Joseph Larmor, ed., *The Scientific Writings of the Late George Francis Fitzgerald* (Dublin: Hodges, Figgis, 1902), xxxii–lxiv, on xxxii. Further on the "Maxwellians," see Buchwald, *Maxwell to Microphysics*, passim, esp. 167–73, where we find that even when Joseph Larmor made his compromise with Continental theory by introducing primitive, microscopic charge carriers, or "electrons," into his own theory, he characterized these as "centres of radial rotational strain in the aether," or "nucle[i] of intrinsic twist" in the ether, and hence still in a sense emergent from the ether; Bruce J. Hunt, "The Maxwellians," Ph.D. dissertation, Johns Hopkins University, 1984, to appear as *The Maxwellians* (Ithaca, N.Y.: Cornell University Press, in press), also, e.g., idem, "Experimenting on the Ether: Oliver J. Lodge and the Great Whirling Machine," *Historical Studies in the Physical and Biological Sciences, 16* (1986): 111–34. See also, e.g., Stuart M. Feffer, "Arthur Schuster, J. J. Thomson, and the Discovery of the Electron," *Historical Studies in the Physical and Biological Sciences, 20* (1989), 33–61; John L. Heilbron (paper delivered at a conference "Cambridge Mathematical Physics in the Nineteenth Century," Grasmere, England, March 1984), cited in Harman, ed., *Wranglers and Physicists*, 11: "Heilbron drew attention to the decline in vitality in research at the Cavendish in the early 1890s as indicated by numbers of research students and publications, and offered certain intellectual and institutional factors [including protracted adherence to strict Maxwellian principle] as causes."

4   Cf. John L. Heilbron (paper delivered at the annual meeting of the History of Science Society in Madison, Wisconsin, October 1978).

5   See, e.g., Morris Kline, *Mathematical Thought from Ancient to Modern Times* (Oxford University Press, 1972), 1197–203.

6   Arthur Koestler, *The Sleepwalkers: A History of Man's Changing Vision of the Universe* (New York: Macmillan, 1959).

INDEX

Entries such as "Maxwell, James Clerk" and "molecular vortices" furnish representative, rather than complete, information on these broad topics; see, further, independent, more specific entries.

Printed in the United States
By Bookmasters